Harlan's Crops and Man: People, Plants and Their Domestication

Harlan's Crops and Man: People, Plants and Their Domestication

Third Edition

H. Thomas Stalker, Marilyn L. Warburton, and Jack R. Harlan

Copublication by American Society of Agronomy, Inc., Crop Science Society of America, Inc., and John Wiley & Sons, Inc.

Editorial Correspondence:
American Society of Agronomy, Inc.
Crop Science Society of America, Inc.
5585 Guilford Road, Madison, WI 53711-58011, USA

agronomy.org
crops.org

Registered Offices:
John Wiley & Sons, Inc., 111 River Street, Hoboken, NJ 07030, USA

For details of our global editorial offices, customer services, and more information about Wiley products, visit us at www.wiley.com.

Wiley also publishes its books in a variety of electronic formats and by print-on-demand. Some content that appears in standard print versions of this book may not be available in other formats.

Library of Congress Cataloging-in-Publication Data

Names: Stalker, H. T. (Harold Thomas), 1950– author. | Warburton, Marilyn L., author | Harlan, Jack R. (Jack Rodney), author.
Title: Harlan's crops and man : people, plants and their domestication / H. Thomas Stalker, Marilyn L. Warburton, and Jack R. Harlan.
Other titles: Crops and man
Description: Third edition. | Hoboken, NJ : John Wiley & Sons, Inc. ; Madison, WI : American Society of Agronomy, Inc. : Crop Science Society of America, Inc. : Soil Science Society of America, Inc., [2021] | Includes bibliographical references and index.
Identifiers: LCCN 2020026833 | ISBN 9780891186335 (hardback)
Subjects: LCSH: Crops–History. | Agriculture–History.
Classification: LCC SB71 .H3 2021 | DDC 633–dc23
LC record available at https://lccn.loc.gov/2020026833

doi:10.2135/harlancrops

Cover Design: Wiley
Cover Image: © ACSESS

Contents

Preface   *viii*

1 **Prologue: The Golden Age**   *1*
Crop Evolution   *2*
The Hunter-Gatherer Stereotype   *3*
What Do Gatherers Eat?   *11*
Understanding Life Cycles of Plants   *20*
General Botanical Knowledge   *23*
Manipulation of Vegatation   *25*
Food Plants in Ritual and Ceremony   *26*
On Sharing the Bounty   *27*
Population Control and the Aged   *29*
Conclusions   *30*
References   *31*

2 **Views on Agricultural Origins**   *37*
Agriculture as Divine Gift   *37*
Domestication for Religious Reasons   *43*
Domestication by Crowding   *45*
Agriculture as Discovery   *46*
Agriculture by Stress   *49*
Agriculture as an Extension of Gathering   *50*
Domestication by Perception   *53*
A No-Model Model   *56*
Geography of Plant Domestication   *59*
An Ecological Approach   *63*
Conclusions   *73*
References   *73*

3 What Is a Crop? *79*
Definitions *80*
Intermediate States *81*
A Short List of Cultivated Plants *86*
Crops That Feed the World *106*
References *107*

4 What Is a Weed? *109*
Definitions *110*
Intermediate States *113*
Crop–Weed Complexes *116*
Some Weed Adaptations *120*
Weeds and History *122*
Conclusions *127*
References *127*

5 Classification of Cultivated Plants *131*
Botanical Descriptions and Names *132*
Problems of Formal Taxonomy *134*
The Gene Pool System *136*
Evolutionary Implications *143*
Conclusions *145*
References *145*

6 The Dynamics of Domestication *147*
Domestication of Seed Crops *147*
Domestication of Vegetatively Reproduced Crops *163*
Conclusions *167*
References *167*

7 Space, Time, and Variation *171*
Kinds of Patterns of Variation *171*
Noncentric Crops *175*
Diffuse Origins *178*
Microcenters *180*
Landrace Populations *181*
Implications for Plant Breeding *183*
Conclusions *190*
References *190*

8 **The Near East**   *195*
Introduction   *196*
Archaeological Prelude   *200*
A Note About Dating Archaeological Sites   *202*
Archaeological Sequence of Village Sites   *204*
Spread of Agriculture Out of the Nuclear Area   *210*
Recorded History   *211`*
Conclusions   *212*
References   *213*

9 **Indigenous African Agriculture**   *216*
Introduction   *217*
Archaeological Prelude   *217*
A Savanna Complex   *223*
Crop Competition and Distribution   *227*
Recorded History   *228*
Décrue Agriculture   *230*
Conclusions   *232*
References   *233*

10 **The Far East**   *236*
Archaeological Prelude   *237*
Recorded History   *240*
Far Eastern Crops   *241*
Hunter-Gatherers of Japan   *257*
Plant Domestication in India   *258*
Conclusions   *259*
References   *259*

11 **The Americas**   *263*
Archaeology   *263*
The Crops   *269*
Indigenous Americans as Biochemists   *283*
Conclusions   *287*
References   *288*

12 **Epilogue: Who's in Charge Here?**   *295*
References   *302*

Preface

Third Edition, 2021

Since the second edition of *Crops and Man* was published some 30 years ago, germplasm collections have expanded greatly, molecular genetics has taken root and is being used to answer age-old questions, and archaeological research has discovered many ancient plant and animal remains, uncovered new sites, and expanded our knowledge of the movement of man and his crops throughout the world. Many of the early studies are no longer possible to continue because hunter-gatherers have all but disappeared except in a few relatively isolated regions.

Crop plant evolution involves an understanding of human behavior, as well as extensive knowledge about plants, what happens to plants as man selects traits that he values, and the importance of these plants in varying societies. The process of evolution takes place over both time and space, and as Jack Harlan so eloquently points out, there is no one model or answer to all questions. In this edition, we made every effort to maintain the basic structure of the previous volumes, while updating information that has evolved during the past 30 years. Most of the original references are still used because evolution of particular plants and many theories have not changed, and the older literature presents the foundation for current work.

Jack Harlan did not formulate his theories and concepts by sitting in an office or library and daydreaming; he explored many regions of the world's centers of diversity. He collected more than 12,000 accessions of cereals, forages, legumes, trees, and fruits from more than 45 countries. Many of these have been extensively used as the sources for disease and insect resistances and to introduce genetic variability to modern production agriculture. He made taxonomic revisions of the genera *Cynodon* and *Sorghum* and studied the evolution of many other species, especially the cereals. He was also involved in archaeological research and had firsthand knowledge of ancient plant types.

Dr. Harlan formulated five concepts as related to crop plant evolution: first, the "Compilospecies" concept where related species intermate to form hybrid swarms with high levels of fitness and aggression, and which are able to expand their ecological range. Secondly, he understood the relationships between crops and companion weeds, and the importance of introgression to maintain diversity in a species. Third, Vavilov's Centers of Origins, which were more centers of diversity than origin, were revised into larger areas. Dr. Harlan recognized that not all crops had distinct centers and that the center of origin is not necessarily (and is more often not), the center of diversity. Fourth, he understood that the origin of crop domestication occurred for different reasons by various peoples and no one concept fits all situations. Thus, he developed a no-model model to incorporate the array of theories for crop domestication. Lastly, a natural classification of cultivated plants was proposed that consisted of gene pools rather than the classical method of morphological descriptions. This allows the thousands of variants of a crop to be lumped together into a single genetically and reproductively unified gene pool.

For his masterful accomplishments and service to the agriculture community, Dr. Harlan received many highly prestigious recognitions and awards, both nationally and internationally. His contributions have been recognized in symposia and in Europe a conference series named after him continues to bring together scientists to discuss topics in crop plant evolution.

Jack Harlan was a brilliant scientist and a true scholar. He stimulated all those who knew him to explore new avenues of learning and to never stop acquiring knowledge, not only in their specialty, but in related fields as well. Jack R. Harlan was my mentor, graduate advisor, and friend.

Harlan's use of the word "man" to describe all people was commonplace at the time of his writing. We have left this gender non-descript word use in our attempt to maintain the original flavor of his entertaining story style, and trust our readers understand we mean no disrespect.

In this revision, we hope that young plant scientists will broaden their views of the world around them to better understand the evolution of humans and the plants that feed the world. The book does not present the genetics of speciation, polyploidy, or plant breeding. But rather, it is intended to present views of evolution through the personal experiences of Jack Harlan and set the foundations for patterns of crop diversity.

H. Thomas Stalker
Raleigh, North Carolina

1

Prologue

The Golden Age

First of all the immortals who dwell on Olympian homes brought into being the golden race of immortal men. These belonged to the time when Kronos ruled over heaven, and they lived like gods without care in their hearts, free and apart from labor and misery. Nor was the terror of old age on them, but always with youthful hands and feet they took their delight in festive pleasures apart from all evil; and they died as if going to sleep. Every good thing was theirs to enjoy: the grain-giving earth produced her fruits spontaneously, abundantly, freely; and they in complete satisfaction lived off their fields without any cares in blessed abundance.

Hesiod, eighth century BC
(Translated by R. M. Frazer, 1983)

Harlan's Crops and Man: People, Plants and Their Domestication, Third Edition.
H. Thomas Stalker, Marilyn L. Warburton, and Jack R. Harlan.
© 2021 American Society of Agronomy, Inc. and Crop Science Society of America, Inc.
Published 2021 by John Wiley & Sons, Inc.
doi:10.2135/harlancrops

Crop Evolution

In this book, we shall be dealing with evolution. We shall try to describe the evolution of crop plants from their wild progenitors to fully domesticated races and the emergence of agricultural economies from preagricultural ones. We shall deal with the activities of man that shaped the evolution of crops and that influenced the shaping of crops as human societies evolved. Crops are artifacts made and molded by man as much as a flint arrowhead, a stone axe head, or a clay pot. On the other hand, man has become so utterly dependent on the plants he grows for food that, in a sense, the plants have "domesticated" *him*. A fully domesticated plant cannot survive without the aid of man, but only a minute fraction of the human population could survive without cultivated plants. Crops and man are mutually dependent and we shall attempt to describe how this intimate symbiosis evolved.

The word *"evolution"* means an opening out, an unfolding, a realization of potential as in the opening of a flower or the germination of a seed. It implies a gradual process rather than sudden or cataclysmic events, with each living thing being derived genetically from preceding living things. Evolution as a process means change with time and the changes may be relatively slow or rapid, the time relatively long or short. Thus, the differences brought about by evolution over time may be small or great. As we shall see, some cultivated plants differ very little, if at all, from their progenitors. The same can be said for the evolution of agricultural economies and the sociological changes that have occurred in the process of developing fully agricultural and industrial societies from hunting–gathering systems.

To develop a degree of understanding of what has happened and what agricultural systems mean to mankind, we need some sort of picture of what life was like before agriculture. We need to establish a baseline from which we can visualize the domestication of plants and the emergence of agriculture. What kinds of plants did man eat before today's crops were available? What did he know about plants, and what might have caused him to begin the process of domestication? The descriptions given here will necessarily be brief and sketchy, but will give an idea of the condition of man before he began to grow plants with the purpose of using them for food.

We also need to know something about man as a hunter to understand ourselves. Lee and DeVore (1968) have put it succinctly:

> Cultural Man has been on earth for some 2,000,000 years; for over 99% of this period he has lived as a hunter-gatherer. Only in this last 10,000 years has man begun to domesticate plants and animals, to

use metals and to harness energy sources other than the human body.... Of the estimated 80,000,000,000 men who have ever lived out a life span on earth, over 90% have lived as hunters and gatherers; about 6% have lived by agriculture and the remaining few percent have lived in industrial societies. To date, the hunting way of life has been the most successful and persistent adaptation man has ever achieved.

As a matter of general education and self-understanding, it is important that we know something about this basic human adaptation. There are two general approaches to the problem: (a) we can study surviving nonagricultural societies and examine the ethnographic observations made within the last few centuries, or (b) we can attempt to interpret preagricultural life from the artifacts, refuse, and other clues left by ancient man and recovered by archaeological techniques. In this chapter, we shall deal primarily with the first approach but the archaeological record shall be touched on in later sections.

The Hunter-Gatherer Stereotype

Traditionally, agricultural people have looked down on hunting people who are described as "savage," "backward," "primitive," "ignorant," "indolent," "lazy," "wild," and "lacking in intelligence." Europeans applied the term "civilized tribes" to some eastern North American natives who lived in towns and cultivated plants, but these Native Americans themselves referred to the hunting tribes of the plains as "wild Indians." In Africa, farming groups that surround hunter-gatherers, ". . . did not merely assert their political dominance over the hunter-gatherers and ex-hunter-gatherers they encapsulated; they also treated them as inferiors, as people apart, stigmatized them and discriminated against them" (Woodburn, 1988, p. 37). Similar attitudes prevail in Asia, Oceania, and Tropical America. The prejudice is nearly universal.

The stereotype includes the idea that hunting–gathering people were always on the verge of starvation and that the pursuit of food took so much of their time and energy that there was not enough of either one left over to build more "advanced" cultures. Hunters were too nomadic to cultivate plants and too ignorant or unintelligent to understand the life cycles of plants. The idea of sowing or planting had never occurred to them and they lacked the intelligence to conceive of it. Hunters were concerned with animals and had no interest in plants. In the stereotype that developed, it was generally agreed that the life of the hunter-gatherer was "nasty, brutish, and short," and that any study of such people would only reveal that they lived like animals, were of low intelligence, and were intellectually insensitive and incapable of "improvement."

Occasionally, an unusually perceptive student of mankind tried to point out that hunting man might be as intelligent as anyone else; that he had a sensitive spiritual and religious outlook; that he was capable of high art; that his mythologies were worthy of serious consideration; and that he was, in fact, as one of us and belonged to the same species with all its weaknesses and potentialities. Such opinions were seldom taken very seriously until recent years. It has finally become apparent that no part of the stereotype is correct and that widely held presuppositions are all completely false and untenable. Our ancestors were not as stupid or as brutish as we wanted to believe.

In 1966, Richard B. Lee and Irven DeVore organized a symposium on Man the Hunter held at the University of Chicago and published in 1968. Lee reported on his studies of the San !Kung of the Dobe area, Botswana. Over a three-week period, Lee (1968) found that !Kung Bushmen spent 2.3, 1.9, and 3.2 days for the first, second, and third week, respectively, in subsistence activities. He wrote, "In all, the adults of the Dobe camp worked about 2 ½ days a week. Since the average working day was about 6 hr long, the fact emerges that !Kung Bushmen of Dobe, despite their harsh environment, devote from 12 to 19 hr a week to getting food."

Among the Bushmen, neither the children nor the aged are pressed into service. Children can help if they wish, but are not expected to contribute regularly to the work force until they are married. The aged are respected for their knowledge, experience, and legendary lore; and are cared for even when blind or lame or unable to contribute to the food-gathering activities. Neither nonproductive children nor the aged are considered a burden.

To the !Kung Bushman, the mongongo nut [*Schinziophyton rautanenii* (Schinz) Radcl.-Sm] is basically the staff of life. These nuts are available year-round and are remarkably nutritious (Table 1.1). The average daily per-capita consumption of 300 nuts weighs "only about 7.5 ounces (212.6 g) but contains the caloric equivalent of 2.5 pounds (1134 g) of cooked rice and the protein equivalent of 14 ounces (397 g) of lean beef" (Lee, 1968). Lee found the diet adequate, starvation unknown, the general health good, and longevity about as good as in modern industrial societies. The average of 2140 calories per person daily (Table 1.1) compares favorably to the 2015 USDA recommendations of 2400–3000 calories for an adult male and 1800–2400 calories for an adult female (https://health.gov/dietaryguidelines/2015/guidelines/appendix-2/).

Table 1.1 Diet of the !Kung Bushmen.

	Protein (g/day)	Calories per person per day	Percent caloric contribution of meat and vegetables
Meat	34.5	690	33
Mongongo nuts	56.7	1,260	67
Other vegetable foods	1.9	190	
Total	93.1	2,140	100

Source: Adapted from Lee (1968).

Sahlins (1968) came in with almost identical figures for subsistence activities of the Australian Aborigines he studied and elaborated on his term "original affluent society." One can be affluent, he said, either by having a great deal or by not wanting much. If one is consistently on the move and must carry all one's possessions, one does not want much. The Aborigines also appeared to be well fed and healthy, and enjoyed a great deal of leisure time.

Gatherers can obtain food in abundance even in the deserts of Australia and the Kalahari Desert of Africa. The rhythm of food-getting activities is almost identical between the Australian Aborigine and the !Kung Bushmen of southern Africa. The women and children are primarily involved in obtaining plant and small animal materials. Hunting is reserved for males at the age of puberty or older but is more of a sport than a necessity. Meat is a welcome addition to a rather dull diet but is seldom required in any abundance for adequate nutrition. Both males and females tend to work for 2 days and every third day is a holiday (Figure 1.1). Even during the days they work, only about 3–4 hr per day are employed to supply food for the entire group (Australian data presented by Sahlins, 1968).

Other reports at the symposium tended to support these general findings. A picture emerged of leisure, if not affluent societies, where the food supply was assured even under difficult environmental conditions and could be obtained from natural sources with little effort. The picture described did seem to fit the golden age of Hesiod or the Biblical Garden of Eden.

The publication of *Man the Hunter* was a surprise to many who believed some version of the hunter stereotype. The stimulation was enormous. Between 1968 and 1992, there were at least 12 international conferences on hunter-gatherers as a direct result, but not all were published. A few of the early conferences included ones published by Ingold et al. (1988a, 1988b) and by Schire (1984). In addition, one may cite Bicchieri

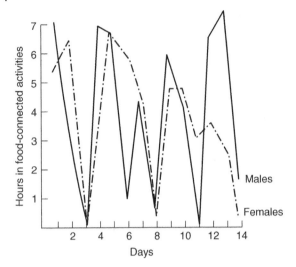

Figure 1.1   Food-gathering activities of the Australian Aborigines. *Source:* Adapted from Sahlins (1968).

(1972), *Hunters and Gatherers Today*; Dahlberg (1981), *Woman, the Gatherer*; Winterhalder and Smith (1981), *Hunter-gatherer Foraging Strategies*; Williams and Hunn (1982), *Resource Managers: North American and Australian Hunter-gatherers*; Koyama and Thomas (1982), *Affluent Foragers: Pacific Coasts East and West*; Price and Brown (1985), *Prehistoric Hunter-gatherers: The Emergence of Social and Cultural Complexity*; Harris and Hillman (1989), *Foraging and Farming: The Evolution of Plant Exploitation*; and such regional treatments as Hallam (1975), *Fire and Hearth: A Study of Aboriginal Usage and European Usurpation in Southwestern Australia*; Silberbauer (1981, p. 242), *Hunter and Habitat in the Central Kalahari Desert*; Riches (1982), *Northern Nomadic Hunter-gatherers*; Lee (1984), *The Dobe!Kung*; Akazawa and Aikens (1986), *Prehistoric Hunter-gatherers in Japan*; and there are many hundreds of additional research papers. There is now a vast amount of new material on the subject, but some of the oldest papers are still the most useful because observations were made before the hunter-gatherers were so restricted and encapsulated as they are now.

The biases of some of the investigators were often clear. Some set out to dispute the "affluent society" concept and others to support it. Some of the anthropologists were hung up on Marxist views of "history," since the egalitarian nature of most hunter-gatherer societies suggested Marx's view of communism: "No one starves unless all starve"; "no man need go hungry while another eats"; "rich and poor perish together," and so forth

(Lee, 1988). The quotes are from observers of Iroquois, Ainu, and Nuer, respectively, and seem to equate egalitarianism with hunger, which is probably not fair. Incidentally, Karl Marx took his model of basic communism from an agricultural Iroquois society, not from hunter-gatherers, who are not so likely to starve.

What do the new studies show? To no one's surprise, they show that the golden age was more golden for some than for others. Even a few examples of famine were found (Johnson & Earle, 1987, p. 374). Brian Hayden (1981) listed a number of tribes showing a continuum of work from "a few minutes per day" (Tanaina in Alaska) or 2 hr per day (Hadza in Tanzania) to "all day every day" or "too busy to visit relatives" (Birhor in India). Well, I have been too busy to visit relatives even when I wasn't doing much of anything. It also comes as no surprise that if processing and cooking time is added to collecting time, it takes longer to get a meal than some figures would suggest. Processing some foods is laborious and time-consuming. Grinding or pounding seeds into flour has always been drudgery, and boiling toxic foods in several changes of water takes a lot of time. Still, is watching a pot boil hard labor, especially if the kids make a game of picking up sticks to keep the fire going? And, of course, farmers must also process their food, too, so the addition of processing and cooking time does not necessarily change the comparison.

There are certain aspects of time and work that do not seem to receive due attention. Suppose you like your work? I always have, and have spent far more time at it than necessary for survival. Consider those men of industrial societies who spend endless hours cramped and freezing in a duck blind for little or no reward, or those who huddle in a shelter fishing through the ice in the middle of a Minnesota winter. The social aspects are what matter; after a few nips of whiskey, no one cares if the rod bends or not. I record two ethnographic notes from my own experience, both from farming societies, but the principles apply to anyone. Early one morning on a deserted road in Afghanistan, I came across a line of men dressed in colorful embroidered jackets, balloon pants, and pixie-toed shoes. They had two drums and were singing and dancing up and down with their sickles in the air. A group of women followed, shrouded in their chadors, but obviously enjoying the occasion. I stopped and asked in broken Farsee: "Is this a wedding celebration or something?" They looked surprised and said: "No, nothing. We are just going out to cut wheat." Harvest time is a good time of year even if it is hot and the "work" is hard. It is a time for socializing and, if the harvest is good, for celebrating.

A second observation was in eastern Turkey. My interpreter and I had seen a family harvesting a field and we stopped. He talked to the people while I collected some samples. My interpreter later told me that he had commented to the farmer that he could harvest the field in half the time if he would use a scythe and cradle. The farmer looked at him in astonishment and said: "Then what would I do?" There is a certain amount of Parkinson's law in all these activities. One fills up the time available. What is the meaning of time if there is more of it than you know how to use? As for getting by with the least effort possible for survival, I do not think that is human nature. Sure, anyone can drink vin ordinaire, but why not work a little harder and drink Chateauneuf-du-Pape?

How do hunter-gatherers spend their leisure? Apparently they sleep a lot, but there are other diversions. Gambling is popular among many tribes; Woodburn (1970, p. 59) states that the Hadza people spend more time in gambling than in obtaining food. The most popular gambling stake is poisoned arrows. There are also music, dances, ritual and ceremony, rites of passage, playing cat's cradle, storytelling, creative arts, making useful and decorative articles, and similar activities. Life appears easy, but generally dull. Perhaps as a consequence there is a great deal of coming and going; the camp population is fluid and camps may be moved on the slightest pretext or for no reason at all. Understandably, there is a tendency to concentrate on the foods most easily obtained at a given time, and these are likely to change from season to season and, to some extent, from year to year. Groups of people in many gathering societies tend to be very fluid for that reason. When food is at maximum abundance, there is a tendency to gather in large bands. This is the season for rejoicing, celebrating, observing ancient tribal rituals, arranging marriages, and having naming ceremonies, coming of age ceremonies, and so on. The tribe is more fully represented at this time. During the most difficult season of the year, the people may break up into microbands to better exploit the gathering range and to avoid exhausting the food supply near the larger camps.

Many Australian Aborigines remain apart much of the yearly cycle even after becoming dependent on European agricultural–industrial systems. For most of the year they find jobs as ranch hands, laborers, mechanics, and so forth, but they may quit whatever they are doing, take off their store-bought clothes, and take a three-month "walkabout" during their traditionally festive season. Gathering is still easier than working at that time of year.

The study of hunting tribes that have survived long enough to have been observed by modern ethnographers is full of difficulties and pitfalls. Many tribes had become profoundly modified through contact with and by the pressures applied by agriculturalists. Some were reduced to the status of slaves or servants; others were restricted on reservations or their normal ranges were constricted by pressures of stronger groups. The social and economic structures of many tribes were in an advanced stage of disintegration at the time of ethnographic description.

The geographic distribution of surviving hunters results in a serious bias. By and large, hunters have survived where agriculture is unrewarding. We find them in the Kalahari Desert and adjacent dry savanna in southern Africa, in small pockets of tropical rain forest, in the frozen wastes of the Arctic, or in western North America, but there are no examples left in the more productive agricultural lands of the world.

At the time of European contact, the eastern forests and woodlands of North America were largely populated by native agriculturalists; the people living in the plains and westward mostly maintained hunting–gathering economies. There were enclaves of farmers, such as the Mandan on the Missouri River in North Dakota, and a highly sophisticated agriculture had developed in the Southwest USA where people practiced irrigation on a large scale and often lived in towns. Some farming was practiced along the Colorado River watershed and into southern California, but most of the California natives and other tribes of western North America lived by hunting, fishing, and gathering. A substantial body of information has been assembled about them, but we must remember that they did have contact with farming people and some of their cultural elements could have been borrowed.

Data for hunter-gatherers in South America have been accumulating during the late 20th and into the 21st centuries. In the review by Scheinsohn (2003), she indicates distinct areas occupied by hunter-gatherers in the grasslands of Argentina and southern Chile, farming communities in the highlands of western South America, and mixed hunter-gatherer and farming societies in more mid-to-low land areas of Bolivia, Brazil, and Venezuela by about 6000 BP (Before Present). There is some evidence of man in South America by at least 30,000 BP (Scheinsohn, 2003), and these peoples were certainly hunter-gatherers. The Bushman of southern Africa has been studied in some detail, but we know historically that they had long contact with the livestock-herding Hottentot and farming Bantu tribes. The Congo pygmies often spend part of each year with agricultural people. The Ainu of Japan have taken up some farming in the last century or so. Many of the hunter-gatherers of India are so constricted by agriculturalists that they have virtually become members of a nonfarming caste.

The Andaman Islanders succeeded in preserving a greater degree of isolation, partly by killing off strangers who landed or were shipwrecked on their shores. Still, we know they borrowed some customs from outsiders. Both pottery and pigs seem to have been introduced about 1500 AD

(Coon, 1971). It is even possible that they were agriculturalists when they arrived and abandoned the practice when they found it unnecessary.

Perhaps our most reliable data come from Australia. At the time of European contact in the early 19th century, there was an entire continent populated by an estimated 300,000 people without a single domesticated plant and no genuine agriculture. Although it is true that for some centuries before European contact there were Malayan traders visiting northern Australia on a fairly regular basis, there is little evidence that this resulted in significant changes in use of food resources and it did not induce the Aborigines to take up the cultivation of plants. The Torres Strait is also rather narrow and some contact with agricultural Melanesians occurred. That this would influence the whole of Australia very much seems doubtful.

I shall, therefore, rely more on ethnographic data from Australia than elsewhere, but will remind the reader that any reconstruction of a way of life of some thousands of years ago, based on a small, biased sample of living people, is full of hazards and sources of error. The earlier accounts may have more value than some of the later ones because the effects of European contact were rapid and profound.

Woodburn (1988) and in a series of papers, outlined an important distinction between immediate return strategies and delayed return strategies. The former live from day to day, or at most a few days at a time on current returns. Delayed return groups have longer-term goals; these include manufacturing of boats, nets, weirs, traps, and deadfalls, tending bee hives, the capture and keeping of animals to be eaten later, the replacement of the tops of yams at digging time, sowing of seeds, managing vegetation with fire, water spreading, irrigation, flooding of forests, arranged marriages, and so forth. The Australian Aborigines were delayed return strategists of great skill, and as such were closer to agriculturalists than to immediate return hunter-gatherers such as the Bushmen and Hadza. Great Basin and West Coast Native Americans and the Jomon of Japan were also delayed return strategists.

As more and more data have accumulated, a consensus has developed that present day and recent hunter-gatherers, whether of immediate or delayed return, have evolved in parallel with agriculturalists and no longer represent the original condition before agriculture. They are not the "pristine" hunger–gatherers of 10,000–12,000 years ago. In addition, the diversity among hunter-gatherers is such that no single model can represent them. There is not even a single model for Australia, let alone the other hunter-gatherers in the world. Our extensive field studies will not tell us all we want to know about preagricultural societies, but they are suggestive.

The oldest remains of *Homo sapiens* L. were left in Morocco about 315,000 years ago (Hublin et al., 2017a, 2017b), which is much older than previously thought. Foley (1988) reserved the term "human" for anatomically modern man who appeared on earth as early as 100,000 years ago and as late as 30,000 years ago in some regions, but many intermediate fossil remains define the evolution within the genus *Homo*. However, early species of *Homo* were not "human." Late Pleistocene man was anatomically modern, but larger, heavier, and more sexually dimorphic. Foley suggests reduction in size and dimorphism was a response to a change in food procurement systems. With the extinction of many large mammals and general impoverishment of the fauna at the end of Pleistocene, men and women began to share more evenly in food procurement, and the broader spectrum of plants and animals exploited was accompanied by morphological changes in humans.

What Do Gatherers Eat?

Lee (1968) classified 58 tribes according to the percentage of dependence on hunting, fishing, or gathering. The data were taken from the Ethnographic Atlas (Murdock, 1967), but adjusted somewhat by transferring the pursuit of large sea mammals from fishing to hunting and shell-fishing from fishing to gathering. The food obtained by gathering is predominantly of plant origin. The class does include small animal foods such as mice, rats, lizards, eggs, insect grubs, and snails. Tortoise and shell-fishing is important to a few gathering tribes. In several cases where detailed analyses were made, however, plant foods contributed 60%–80% of the intake of gathering people.

In his *List of Foods Used in Africa*, Jardin (1967) compiled an extensive and complex list of species. I have attempted to remove cultivated plants and introductions and reduce the synonymy as much as possible. There still remain more than 1,400 species that could be grouped into classes as follows:

Grass seeds approximately 60 spp.
Legumes approximately 50 spp.
Roots and tubers approximately 90 spp.
Oil seeds approximately 60 spp.
Fruits and nuts >550 spp.
Vegetables and spices >600 spp.
Total >1410 spp.

Most of Jardin's reports concerned agricultural tribes and only a small fraction of the list represented foods of gatherers. This suggests that (a) many more species have been gathered from the wild than have ever been domesticated, (b) even after agriculture is fully developed, gathering wild plant foods is still a worthwhile effort, and (c) wild plant resources are of the same general kinds as domesticated plant resources. See also Fox and Young (1982) for southern Africa.

Yanovsky (1936) in his *Food Plants of the North American Indians* lists 1112 species of 444 genera belonging to 120 families. About 10% of these are crops or imported weeds; the rest are native American plants. The bulk of the plants listed were gathered by nonagricultural tribes. Fernald and Kinsey (1943) listed about 1000 species for eastern North America alone. Plants gathered in Central and South America have not been conveniently compiled, but the number of species is very large. A partial listing is given by Lévi-Strauss (1950) in *The Use of Wild Plants in Tropical South America*.

Our most reliable information again might come from Australian areas where agriculture was not practiced and where none of the plants had been domesticated. Lists compiled by Cribb and Cribb (1975), Irvine (1957), Levitt (1981), and Maiden (1889), are of help here, although no list is complete; there are problems of identification and synonymy, and many of the early ethnographic records contain native names because the observers were not botanists and could not identify the plants. Even so, Australians were recorded as having gathered and used over 400 species belonging to 250 or more genera.

Some observations are grouped below according to general kinds of plant food resources.

Grass Seeds (Potential Cereals)

Seeds of wild grasses have long been an important source of food and are still harvested on a large scale in some regions. A.C. Gregory (1886) commented:

> On Cooper's Creek (Australia), the natives reap a Panicum grass. Fields of 1000 acres (405 ha) are there met with growing this cereal. The natives cut it down by means of stone knives, cutting down the stalk half way, beat out the seed, leaving the straw which is often met with in large heaps; they winnow by tossing seed and husk in the air, the wind carrying away the husks. The grinding into meal is done by means of two stones—a large irregular slab and a small cannon-ball-like one; the seed is laid on the former and ground, sometimes dry and at others with water into a meal.

Stickney (1896) described methods of the wild rice (*Zizania palustris* L.) harvest by the Ojibwa of Wisconsin late in the 19th century:

Two women, working together in a canoe, took a large ball of cedar bark twine and tied up sheaves just below the panicles when the seed was in the milk stage. Later, they went back when the seed was ripe and beat the sheaves over the canoe. Each woman knew her own bundles and the right of ownership was scrupulously respected. Sometimes sheaves were not previously prepared and the woman in the back would pole slowly forward while the other reached out with a curved stick and bent a bunch of stalks over the canoe and hit them with a straight stick held in the other hand. About a gill is attached at each blow. When the canoe became heavily laden in the front, the women exchanged implements as they kept their places and the canoe was poled back in the opposite direction. When the canoe was fully loaded and low in the water it was beached and the wild rice removed. The wild rice was dried in the sun or on a platform over a fire. Dehulling was done by men who placed the seed in a skin bag and treaded it in a pit dug in the soil. Dehulled seed was stored in bark boxes or large skin bags; sometimes so much seed was stored that it lasted until the next harvest.

Wild races of common Asian rice (*Oryza sativa* L.) were once harvested on a considerable scale in northern Australia (Bancroft, 1884):

> The wild rice of the Carpenteria swamps (*Oryza sativa*), however, needs to be carefully cleaned from its spiny chaff, which may be done by rubbing in wooden troughs. This must be the most important grass-food in Australia, being little inferior to cultivated grain. The plant grows six feet (1.8 m) high, and produces a good crop even in the latitude of Brisbane. The "paddy" is black with long awns. It is interesting, in Australia, to find one of the original sources of a cereal that has been cultivated in Asia for thousands of years.

The wild races are still harvested in India despite the cultivation of domesticated forms for six or seven millennia (Roy, 1921):

> In the Central Provinces the Gonds and Dhimars harvest this rice by tying the plants together into clumps and thus preventing the grains from falling. These grains have also got a certain demand in the market as they are often used by devout Hindus in these parts on fast days besides being sold to the poorer classes.

Burkill (1935) makes a similar observation:

The poor do not ignore it (wild rice), but tying the awns together before maturity save the grain for themselves, or they collect the fallen grain, which is made an easier process by the length of the awns.

Ping-Ti Ho (1969) documented the harvesting of wild rice over much of southern and central China during a span of an entire millennium. One report, dated 874 AD, from Ts'angchou, Hopei Province, to the emperor may be paraphrased: "Wild rice ripened in an area of more than 200,000 mu (13,000 ha), much to the benefit of the poor of local and neighboring counties" (Ho, 1969). It is to be noted that rice had been a major crop in China for over 6000 years at the date of this report, but that the gathering of seeds of wild rice was still worth the effort.

I have observed other species of rice, *O. barthii* A. Chev. and *O. longistaminata* A. Chev. & Roehr., that are regularly harvested in Africa, sometimes in sufficient abundance to appear in the markets. The Africans sometimes also tie wild rice into clumps before harvest (Harlan, 1989). Claude Lévi-Strauss (1950) reports the harvesting of *O. subulata* Nees [syn. *Rhynchoryza subulata* (Nees) Baill.] in Uruguay, Rio Grande do Sul, and the marshes of the upper Paraguay and Guaporé Rivers in South America. He also reports the technique of binding before harvest:

The Tupí-Cawahíb of the upper Madeira River gather the seeds of an unidentified wild grass that grows in the forest, and to facilitate the harvest they tie together several stems before they are ripe, so that the seeds of several plants fall on the same spot and pile up in small heaps.

Panicum has been a favorite grass seed of gatherers the world over. In North America, *P. capillare* L., *P. obtusum* Kunth [syn. *Hopia obtusa* (Kunth) Zuloaga & Morrone], and *P. urvilleanum* Kunth have been listed as harvested in the wild (Yanovsky, 1936), and *P. hirticaule* J. Presl var (syn. *P. sonorum* Beal) was domesticated in Mexico (Gentry, 1942; Nabhan & deWet, 1984). Seven species are listed for Africa (Jardin, 1967), with the most important being *P. laetum* Kunth and *P. turgidum* Forssk. Four species are recorded for Australia, with *P. decompositum* R. Br. occurring in 1000-ha fields. Two species, *P. miliaceum* L. and *P. antidotale* Retz. were domesticated in Eurasia and India, respectively. It appears that food gatherers are attracted to similar plants.

At least five wild species of *Sporobolus* were harvested in North America, three in Africa, and three in Australia. Species of *Eragrostis* were gathered in North America, Australia, and Africa. For Africa, six wild species are listed and one was domesticated as a cereal, *Eragrostis tef* (*Zuccagni*) Trotter in Ethopia. *Eleusine* and *Dactyloctenium* were harvested in Australia, India, and Africa with one species (*E. coracana L. Gaertn.*) being

Table 1.2   Analysis of wild and cultivated wheats.

	Ether extract (%)	Crude fiber (%)	Crude protein (%)	NFE (%)[a]
Wild einkorn	2.64	2.33	22.83	60.04
Modern wheat	1.50	1.33	10.79	75.01

Source: Adapted from Harlan (1967).
[a]Nitrogen-free extract or carbohydrates other than fiber.

domesticated. Species of Digitaria were harvested in Australia, India, Africa, and Europe. Digitaria exilis (Kippist) Stapf and D. iburua Stapf were domesticated in Africa, D. cruciate Nees ex Hoff. f. in India, and common crabgrass [D. sanguinalis (L.) Scop.] was cultivated as a cereal in central Europe until the 19th century without actually being domesticated (Körnicke, 1985). The differences between cultivation and domestication will be discussed in Chapter 3.

Mannagrass [*Glyceria fluitans* (L.) R. Br.] was harvested in substantial quantities from the marshes of central and eastern Europe as late as 1925 (Szafer, 1966). The seed was even exported from the port of Danzig to countries around the Baltic. Yanovsky (1936) reports that the same species was harvested by Native Americans in Utah, Nevada, and Oregon. Wild oats (*Avena barbata* Pott ex Link and *A. fatua* L.) were harvested by the Pormo tribe in California after these weedy plants had been introduced from the Mediterranean (Gifford, 1967). As late as 100 years ago, wild grass seeds were harvested on a commercial scale in central Africa and exported by camel caravans into the desert and other food deficit areas (Harlan, 1989).

I once studied the amount of grain that could be harvested from wild einkorn wheat (*Triticum monococcum* L. subsp. *boeoticum*) in Turkey (Harlan, 1967). I found no difficulty in collecting over 2 kg of head material or the equivalent of 1 kg of clean grain per hour. On analysis, the grain contained about 23% protein as compared to about 11% for modern cultivated wheat (Table 1.2).

In all, Jardin (1967) lists about 60 species of wild grasses that have been harvested for their seeds in Africa within recent decades. Yanovsky (1936) lists approximately 38 for North America, and Irvine (1957) and others mention about 25 for Australia. The exact number cannot be given because of problems with synonyms and identification. Relatively little is known about wild grass harvesting in Europe and Asia although *Oryza, Panicum, Digitaria,* and *Glyceria* have been mentioned.

Legumes (Potential Pulses)

Gathering peoples are evidently attracted to Leguminosae of various kinds. Whole pods may be used, as well as seeds only, pods only, or even the tissues inside the pods surrounding the seeds. Some legumes have edible tubers and others have leaves or young shoots suitable for pot-herbs. Not infrequently the material harvested is poisonous and must be detoxified before use. Poisonous materials can be used for stunning fish, stupefying emus, or making poison arrows.

As with the Gramineae, certain genera appear frequently on plant lists and several distinct species of a given genus may be used in different parts of the world. Genera with wide distributions may be very widely used. For example, many species of Acacia are exploited in Australia, several are used in Africa and Asia, but only a few are used in the Americas. More species of *Prosopis* (mesquite) are used in the Americas, however, than in Africa, Asia, and Australia. Different species of *Canavalia* are harvested in Central and South America and in Southeast Asia and Australia. *Vigna* and *Dolichos* are widely exploited in Africa, Asia, and Australia while several species of *Phaseolus* are harvested in the Americas. *Tephrosia* spp. have been used for fish poisons on five continents.

Root and Tuber Plants

Roots, tubers, rhizomes, and bulbs have been widely harvested for untold millennia. The choice depends more on what is abundant and available than anything else. The genus *Dioscorea* is very large and includes about 600 species distributed throughout the warmer parts of the world. Many produce tubers that are edible or rendered edible after detoxification. About 30 species are harvested in the wild in Africa (Jardin, 1967) and several have been domesticated. Wild yam harvests are important in India, Southeast Asia, the South Pacific, Australia, and tropical America.

Tubers and rhizomes of the Araceae are widely harvested in the tropics and a few are found in the more temperate zones. Bulbs of the Liliaceae are popular where they occur. Yanovsky (1936) lists about 90 species belonging to the lily family (Liliaceae) that supplied food for North American natives. No less than 17 species of wild onion (*Allium*) were listed, and even the death camus *Zygadenus* was eaten after suitable detoxification. Tuberous legumes in the genera *Solanum, Ipomoea, Nymphaea,* and *Eleocharis* have been widely harvested, and *Cyperus rotundus* L. has supplied food in North America, Africa, Asia, Australia, and Europe.

Oil Plants

Most gatherers had periodic access to animal fats, but sources of vegetable oil were also sought. In the wetter tropics, the fruits of various palms (Palmaceae) were especially attractive. The African oil palm (*Elaeis guineensis* Jacq.) is still exploited in the wild as is its counterpart in South America [*E. guineensis* Jacq. (syn. *E. melanococca* Gaertn.)]. Other palms also supply oil in quantity including, of course, the coconut (*Cocos nucifera* L.). Seeds of Compositae, Cruciferae, and Cucurbitaceae are harvested on every continent, partly for their oil content. Many nuts and some fruits are high in oil and are still harvested in the wild. Some familiar ones are *Aleurites* (Candlenut or tung-oil tree), *Persea* (avocado), *Theobroma* (cacao), *Pistacea* (pistachio), *Olea* (olive), and *Butyrospermum* (shea butter tree or karité). Several species of *Sesamum* and *Linum* are harvested for their oily seeds.

Fruits and Nuts

Long lists of fruits and nuts can be compiled, but it is not necessary to go into much detail here. We need only point out that the same patterns prevail as for grass seeds, legumes, and oil plants in that different species of the same genera are exploited almost everywhere they occur. In temperate zones, for example, species of walnut (*Juglans*), hickory (*Carya*), hazelnut (*Corylus*), chestnut (*Castanea*), beech (*Betula*), oak (*Quercus*), hawthorn (*Crataegus*), hackberry (*Celtis*), plum-cherry (*Prunus*), bramblefruits (*Rubus*), grape (*Vitis*), elderberry (*Sambucus*), pine-nuts (*Pinus*), and others were popular with gatherers in Europe, Asia, North America, Africa, and Australia. In the tropics, some of the popular genera were (and are) *Ficus*, *Citrus*, *Musa*, *Syzygium*, *Pandanus*, *Spondias*, *Adansonia*, *Artocarpus*, *Annona*, and *Carica*. If a plant appeals to one gathering tribe, a similar plant is probably used by another tribe, even on another continent.

Vegetables

Because the same general pattern is operative, it might be worthwhile to call attention to repetitive patterns in two families whose produce appeals to gatherers.

Solanaceae. The genus *Solanum* is found on every continent and includes several hundred species. About 15 species are gathered for food in Africa, 9 are listed for North America, and several are found in South America, India, and Australia. Some must be detoxified before being eaten. The fruits are the parts eaten in most cases, but leaves may be used as pot-herbs and a number of species have edible tubers. *Physalis* is another genus widely exploited with at least 10 species gathered in North America plus others in South America, Africa, Europe, Asia, and Australia. Species of wild *Capsicum, Cyphomandra,* and *Lycopersicon* were gathered in the Americas. The genus *Nicotiana* was a favorite of gathering tribes in the Americas and Australia. Several distinct species were involved and they were utilized almost wherever they occurred. In the Americas, the tobaccos were both chewed and smoked, while it was a masticatory only in Australia. Lime of some sort was often mixed with the quid. *Datura* was used as a drug, medicine, or hallucinogen in both eastern and western hemispheres.

Cucurbitaceae. Plants of this family were often attractive to gathering peoples and in some cases were very important because of their abundance. In Australia, Maiden (1889) observed that *Cucumis trigonus* Roxb. was sometimes "growing in such abundance that the whole country seemed strewed with the fruit." In southern Africa, the landscape may be almost cluttered with wild watermelon [*Citrullus colocynthis* (L.) Schrad.] where it may serve as the only source of water for man and animals alike over extended periods of the dry season (Story, 1958). Tropical *Cucumis* and *Mamordica* species are still gathered in the wild in Africa and Asia. The genus *Cucurbita* is confined to the Americas and was extensively exploited by the Native Americans; several species were domesticated. The white-flowered bottle gourd [*Lagenaria siceraria* (Molina) Standl.] has been widely exploited, primarily for the hard shells of the fruits which make excellent containers. Its use has been recorded in the Americas, Africa, Asia, Europe, and Australia, but its distribution as a wild plant is not well known. The fruits of the Australian races are said to be purgative or even poisonous according to Maiden (1889) but are eaten by the Aborigines after being processed. The fruits of some domesticated races may be eaten when young without special precautions. *Luffa* is also widely used in Asia and Africa as a vegetable or medicine, but is a fish poison in Australia (Palmer, 1883).

Summary

Finally, we might return to the plants gathered by Australian Aborigines as, perhaps, representing a most authentic selection by surviving nonagricultural peoples. A short list of genera that include one or more species harvested in the wild by native Australians is given in Table 1.3. I have attempted to indicate where species of each genus are harvested in the wild in addition to Australia. It seems evident from these data and the foregoing discussion that gatherers exploit about the same range of plants wherever they find them.

It is not surprising, therefore, to find independent domestications of different species of the same genus, and if the genus is widespread, the different domesticates may have originated in different continents. Examples of such vicarious domestications occur in the following genera, among others:

1) Mesoamerica and South America—*Amaranthus, Annona, Canavalia, Capsicum, Carica, Chenopodium, Cucurbita, Gossypium, Opuntia, Pachyrrhizus, Phaseolus,* and *Physalis;*

Table 1.3 A short list of genera that include one or more species harvested for food by native Australia.[a]

Acacia (Af, Am, As)+	*Lepidium* (Am, Af, As, E)+
Adansonia (Af, As)+	*Linum* (As, E)+
Aleurites (As, 0)+	*Loranthus*
Alocasia (As, 0)+	*Lucuma* (Am)+
Amamnthus (Am, As, Af)+	*Luffa* (As)+
Amorphophallus (Af. As, 0)+	*Lycium* (Af)
Antidesma (As)	*Macadamia* (#)
Araucana (Am, O, #)	*Manikara* (Af)
Austromyrtus	*Marsilia*
Boerhaavia (Af, Am, As)	*Mimusops* (Af, As)
Bowenia	*Mucuna* (As)+
Calamus	*Musa* (As, 0)+
Canavalia (Af, Am, As)+	*Nasturtium* (As)+
Capparis (Af, As)+	*Nelumbium* (Af, As)+
Carissa (Af)	*Nymphaea* (Am, Af, As)+
Cassia (Af, As, O)	*Ocimum* (Af, As)+
Chenopodium (Am, Af, As)+	*Oryza* (Am, Af, As, O)
Citrus (As, O)+	*Oxalis* (Am, Af, As)+
Clerodendrum (Af)	*Pandanus* (As, O)+
Cordia (Af, As)	*Panicum* (Am, Af, As, E)+
Cucumis (As, Af)+	*Parinari* (Af)
Cyperus (Am, Af. As)	*Phragmites* (Af, Am, As, E, O)
Dactyloctenium (Af)	*Physalis* (Am, Af, As, E)+
Digitaria (Af, As, E)+	*Piper* (Am, Af, As, O)+
Dioscorea (Am, Af. As)+	*Podocarpus* (Af, As, O)+
Diospyros (Am. As, O)+	*Polygonum* (Am, Af, As. E, O)
Dolichos (Af. As. O)+	*Portulaca* (Am, Af. As, E)+
Eleocharis (Am. Af, As)+	*Rubus* (Am, Af, As, E)+
Eleagnus (As, E)+	*Rumex* (Am. Af, As, E)
Eleusine (Af. As)+	*Sambucus* (Am, Af. As, E)+
Eragrostis (Am, Af)+	*Sesbania* (Am, Af. As)+
Eriochloa (Af, As)+	*Solanum* (Am, Af, As, E, O)+
Eucalyptus (#)	*Sorghum* (Af, As)+

(Continued)

Table 1.3 (Continued)

Eugenia (Af, As, O)+	*Spondias* (Am, Af, As, O)+
Ficus (Am, Af, As, O)+	*Sporobolus* (Am, Af)
Gardnia (Af, As, O)+	*Tacca* (Ac, O)+
Gastrodia	*Terminalia* (As, O)+
Geranium (Am, As, E)+	*Trigonella* (As)+
Glycine (As, O)+	*Typha* (Am, Af, As, E, O)+
Grewia (Af, As. O)	*Vigna* (Af, As, O)+
Haemadorum	*Vitex* (Am, Af, As, O)+
Hibiscus (Am, Af, As, O)+	*Vitis* (Am, As, E)+
Ipomoea (Am, Af, As, O)+	*Zamia* (O)
Lagemyia (Am, Af, As, O)+	*Zizyphus* (Af, As)+

[a] Abbreviations in parentheses indicate species harvested in the wild in addition to Australia: Am = America, Af = Africa, As = Asia, E = Europe, O= Oceania, # = modern domestication, + = one or more cultivated species somewhere in the world, but not in Australia.

2) Africa and Asia—*Amorphophallus, Cucumis, Digitaria, Dioscorea, Dolichos, Hibiscus, Oryza, Piper, Solanum,* and *Vigna;*
3) Old and New Worlds—*Amaranthus, Canavalia, Dioscorea, Gossypium, Ipomoea, Lepidium, Lupinus, Panicum, Prunus, Setaria, Solanum,* and *Spondias.*

Understanding Life Cycles of Plants

What do nonagricultural people know of the life cycles of plants? Do they know that flowers lead to seeds and that seeds can be sown to produce more plants? Is this something that must be learned or discovered to commence the domestication of plants or is this a part of the general botanical knowledge of gathering peoples?

A look at the ethnographic evidence shows that some gatherers do plant seeds. Seven of 19 groups studied by Steward (1941) in Nevada sowed seeds of wild plants (Downs, 1964). No tillage was practiced; the seedbed was generally prepared by simply burning the vegetation the previous fall and seeding in the spring. The seeds sown were of entirely wild plants; the most frequently mentioned were species of *Chenopodium, Oryzopsis, Mentzelia,* and *Sophia* (Steward, 1941).

The Paiute tribe of Owens Valley, California, practiced irrigation and also broadcast seeds to thicken up stands of desired plants, but none was domesticated. Irrigation was designed to increase production of wild plants such as *Nicotiana attenuata* Torr. ex S. Watson, *Salvia columbariae* Benth., *Chenopodium fremontii* S. Watson, *Chenopodium album* L., *Helianthus bolanderi* A. Gray, *Eriocoma hymenoidsis* (Roem. & Shult.) Rydb. and *Eleocharis* spp. The earthen dams were simple, but the rather extensive canals required considerable labor to build. One block covered about 5 km² and another close to 13 km² (Steward, 1934). As previously pointed out, however, Natives of the Great Basin could have been influenced by neighboring agriculturalists and their botanical knowledge may not be typical of gatherers in preagricultural times. Let us look elsewhere.

To the Andamanese, the goddess Puluga symbolizes the southwest monsoon that brings violent winds and rains from April to October (Coon, 1971):

> Puluga owned all the wild yams and cicada grubs that the people ate, and all the beeswax that they used in hafting, calking, and cordage. Women who dug yams had to replace the tops to fool Puluga...
>
> Indeed, if Puluga caught the people misusing her property she would get angry and send bad weather. Here we see the practice of planting reinforced by a religious belief. The practice is useful to the people, but does not of itself prove understanding.

An early observation of Sir George Grey (1841) concerning Australian Aborigines is more revealing:

> The natives have, however, a law that no plant bearing seeds is to be dug up after it has flowered; they then call them (for example) the mother of Bohn, the mother of Mud-ja (*Haemadorum* spp.), etc.; and so strict are they in their observance of this rule that I have never seen a native violate it, unless requested by an European, and even then they betray a great dislike to do so.

The practice is confirmed by Gregory (1886):

> The natives on the West Coast of Australia are in the habit among other things of digging up yams as a portion of their means of subsistence; the yams are called "ajuca" in the north and "wirang" in the south. In digging up these yams they invariably re-insert the head of the yams so as to be sure of a future crop, but beyond this they do

absolutely nothing which may be regarded as a tentative step in the direction of cultivating plants for their use.

There seems to be little doubt that the life cycles of plants were well understood by native Australians. The Aborigines were equipped with all the knowledge necessary to practice agriculture, but did not do so.

Klimek (1935) recorded 11 tribes of California peoples that grew a local species of tobacco, but no other crop. Some tribes in Oregon, Washington, and British Columbia followed the same practice (Drucker, 1963). The tobacco was usually either *N. attenuata* or *N. quadrivalvis* Pursh. Harrington (1932) made a very detailed study of tobacco among the Karuk, and found the extent of botanical knowledge remarkable. The Karuk burned logs in the forest and sowed seeds in the ashes. A tobacco garden was called "to plant" or more literally "to put seed." The Karuk had terms for cultivated tobacco, wild tobacco, roots, stems, bark, leaves, branches, leaf branches, pith, gum, flowers, buds, seedpods, flower stem, clusters of flowers, sepals, and calyx. No standard word was used for petal, but descriptive terms were used, for example, the white-flowered *N. quadrivalvis* was said to have "five white ones sticking out." The stamens and pistil were described as "sticking out in the middle of every flower where the seeds are going to be." Stamens are "flower whiskers," "flower threads," or "flower hairs." Pollen is "flower dust." Nine stages of flowering to seed setting were recognized with descriptive terms. There was a classification of seeds, grains, seeds in the midst of a fruit (pit), seeds inside a shell (nut), and so on.

The translation of an informant's description of germinating tobacco seed is botanically accurate and detailed (Harrington, 1932, p. 61):

> Its seeds fall to the ground. The dirt gets over them. Then, after a while, when it gets rained on, the seed sprouts. Sometimes all the seeds do not grow up. They say sometimes some of the seeds get rotten. Its sprouts are small white ones, pretty near the size of a hair. Whenever it is just peeping out, its seed is on top of it. Then they just have two leaves, when they first peep out of the ground. They grow quickly when they grow; in a little while they are tall ones.

The Karuk fertilized with ashes, sowed, weeded, harvested, selected for strength, cured, stored, and sold tobacco, but grew no other crop. Clearly, the concept of planting seeds was in no way revolutionary and did not lead to food production (Harrington, 1932). Where did people ever get the idea that hunter-gatherers did not know about life cycles of plants? The egocentricity of agriculturalists is extraordinary. The situation is summed up succinctly by Flannery (1968): "We know of no human group on earth so primitive that they are ignorant of the connection between plants and the seeds from which they grow. . . ."

General Botanical Knowledge

We should not be surprised if gathering peoples know a lot about plants. They are the real "professional botanists"; for them, life depends on an adequate knowledge of plants. We have seen that gatherers are familiar with hundreds of species and their uses for food. We have noted that many are poisonous and must be detoxified before they can be eaten.

Since "ignorance" is part of the stereotype developed by agricultural people about gatherers, I would like to call attention to an episode described with some apparent pleasure by Sir George Grey (1841). Some of the crew of Captain Cook's expedition of the 1770s observed the Aborigines eating seeds of *Zamia* (a cycad). The crew tried some of their own harvest of *Zamia* and became very ill. They concluded that the Aborigines must have very strong constitutions to be able to live on such food. Later, on shipboard, they fed *Zamia* seeds to some pigs, and a few died. Their admiration for the physical stamina of the natives increased substantially. The Aborigines, of course, had removed the poison before eating their seeds, and were, no doubt, amused at the "ignorance" of their European visitors.

Detoxification is required for a considerable number of plants used by the native people of North American, the Australian Aborigines, and gatherers in tropical zones. Some plants are deadly poisonous without treatment, others only unpleasant. Several acorn species are sweet and need no treatment, while others contain various amounts of tannins. Among California Native Americans, some of the bitterest oaks were the most popular; when properly leached, the original tannin content did not cause any harm. Tribes on the edge of the "acorn belt" were often more selective since they did not depend much on acorns and did not want to go to the trouble of leaching. Leguminous seeds, Solanaceous fruits, *Dioscorea* spp., and Aroid tubers are still among the more common poisonous foods consumed by gatherers.

Detoxification is usually by heat, leaching, or both. The plant material is frequently reduced by grinding or pounding in a mortar to facilitate treatment. Boiling water may be poured through the meal, the material may be boiled in several changes of water, or sometimes prolonged soaking in cold water is enough. Some foods are roasted, pounded, and then leached. Sieves, strainers, cloth sacks, wooden troughs, or sandbeds may be produced for the purpose. Pottery is not necessary; water may be boiled in baskets, hides, wooden boxes, or pits in the ground by dropping fire-heated rocks into the water.

Gatherers not only know how to make poisonous foods safe, but they also know a great deal about drugs, narcotics, medicines, fish poisons, arrow poisons, gums, resins, glues, dyes and paints, bark cloth, and woods for spears, arrows, bows, shields, fire sticks, and canoes. They have also used their botanical knowledge in spinning and weaving, basket-making, and constructing household utensils, fish traps and weirs, masks, figurines, and ceremonial objects.

The Australian Aborigine was fond of chewing a wild tobacco (mostly *Nicotiana suaveolens* Lehm.). Wood of *Acacia salicina* Lindl. was burned to provide ash to mix with the quid. Why this particular species out of dozens of *Acacia?* Johnston and Cleland (1933) analyzed the ash and found it extraordinarily high in calcium sulfate, "sulfuric anhydride 30.09% and lime 40.70%." The alkaloids are more soluble in alkaline solutions. Perhaps any source of lime would do, but the practice reminds one of the custom in India of burning heartwood of *Senegalia catechu* (L. f.) P. J. H. Hunter and Mebb (syn: *Acacia catechu* (L.f.) Wild, Oliv.) to obtain "cutch" which is mixed with other ingredients and used when betel nuts are chewed.

Another masticatory of the Aborigines was *Duboisia hopwoodii* (F. Muell) F. Muell. This is of a different order of drug potency and contains hyoscyamine and norhyoscyamine, with scopolamine in the younger leaves (Johnston & Cleland, 1933). Both narcotics were confounded under the general name "pituri" and were important articles of trade over great distances; "shields, boomerangs, spears and other articles being sent in return for them."

In the late 19th century, Father Trilles, a French missionary to Gabon, West Africa, observed pygmies making arrow poison. The process was long, complex, and dangerous, for the poisons were extremely potent. Ingredients of 10 different plants were used; eight were poisonous and two were gums to be impregnated with poison and stuck to the arrow heads. Two animal poisons were also included: beetle larvae and venom of a horned viper. The procedure is described in *The Hunting Peoples* by Coon (1971) who added this comment:

> A tourist driving along a forest-lined road, seeing an elderly, diminutive black man clad in a bark-cloth breechclout, would have no reason to suspect that this child of nature knew the properties of many medicinal plants, some still undescribed in Western science, and how to combine them for their greatest effect. With the forest

and marsh his pharmacy, his laboratory a secret nook in the shade of tall trees, and a minimum of equipment, the Pygmy poison-maker performs a delicate, dangerous, and highly skilled sequence of operations as exacting as some modern professions.

An indication of ecological sophistication is reported by Levitt (1981) for Aborigines of Groote Eylandt. Some common grasses were used as "calendar plants"—when grains of *Chrysopogon* spp. are ripe, it is time to dig yams; or when grains of *Heteropogon triticerus* R. Br. start to shatter, it is time to dig yams; and when all grains have fallen, it is time to stop. When *Heteropogon contortus* (L.) P. Beauv. ex Roem. & Schult. begins to flower, the rainy season will soon be over. Other hunter-gatherers receive similar signals from their knowledge of plant growth and reproduction.

The more one studies the wealth of plant lore of gathering peoples the more one is impressed by the extent and coverage of their botanical knowledge. Man knows what he needs to know or learns what he must or else he dies. The security and stability of gathering economies are from necessity, rooted in an extensive body of information about plants.

Manipulation of Vegatation

Kangaroo Island lies off the south coast of Australia. It had once been inhabited by Aborigines, but they left or died out long before European contact. The woody vegetation had become a virtually impenetrable thicket, while the nearby mainland with the same climate supported an open, grassy woodland. This comparison gives us some understanding of the extent of Aboriginal control over the vegetation. To this day, if areas are uninhabited for an extended period, the woody vegetation thickens up, and the Aborigines find the landscape uncomfortable and spiritually dangerous (Chase, 1989). After repeated burnings, the land again shows the stamp of human occupancy and the Aborigines feel more comfortable and spiritually safe. The Aborigines have more or less domesticated the landscape by skillful use of fire and complain that the white man lets the land get "dirty" (Lewis, 1989). Jones (1969) called it "firestick farming."

The Aborigines did more than clear land by burning. They diverted water to flood forests in the dry season: "We like to see plenty of water in the jungle all the time, for birds of all kinds gather near it, and the food plants that we like grow better" (Campbell, 1965). They constructed water-spreading devices for the rainy season (Lourandos, 1980), and they ditched to increase the supply of eels and other fish (Walters, 1989). In the course of digging up wild root crops, they churned up large areas to the point they resembled plowed fields. Sir George Grey wrote (1841):

> In the Province of Victoria, as already stated, I have seen tracts of land several square miles in extent, so thickly studded with holes, where the natives have been digging up yams (*Dioscorea*) that it was difficult to walk across it. Again, in the sandy desert country which surrounds for many miles, the town of Perth, in Western Australia, the different species of *Haemadorum* are very plentiful.

The borderline between gathering and farming becomes very hazy at this point. Douglas Yen referred to such activities as Aboriginal agronomy (Yen, 1989). Perhaps the key difference here between foraging and farming is that no native Australian plant was actually domesticated, otherwise hunter-gatherers do about everything farmers do.

The Great Basin Native Americans did about the same, burning vegetation, sowing seeds, and irrigating tracts of land (Downs, 1964). Indeed, fire was used to modify vegetation just about anywhere that vegetation could be burned, and the practice may well have gone back to Acheulean times (Hallam, 1975). There are some immediate returns from the practice; animals fleeing fires are more vulnerable to the spear and the bow, but the major returns are delayed. New shoots, unencumbered by old growth attract grazing animals; the ash provides some fertility for regrowth; heat renders phosphorus more available; woody vegetation is retarded and herbaceous plants increase; wild seed harvests are enhanced; roots and tubers escape injury in the dry season and thrive as competition is reduced. The landscape is tamed, but the plants and animals were not.

Food Plants in Ritual and Ceremony

Some California tribes, heavily dependent on acorns for food, conducted an annual spring ceremony, usually in April, for the purpose of increasing the crop. The participants went out at night, visited specified trees, and implored them to yield abundantly. The trees were supposed to respond (Loeb, 1934).

First-fruit ceremonies are practiced by the African Bushmen. When the fruit of a certain species begins to ripen at the onset of the big rainy season (usually February), a day is appointed and the women go out and ceremonially gather fruit from previously designated trees. The men stay in camp and all the camp fires are extinguished. When the fruits are brought to camp, a composite sample is carefully selected and presented to a head man, who kindles a special fire and ceremonially appeals to the fire for a plentiful harvest. He then eats the fruit. After the ceremony both men and women can partake of the fruits, but it is offense to eat them before the ceremony (Marshall, 1960).

Among various Bushmen tribes at least simple first-fruit ceremonies are performed for a dozen or more different plants. Each of the major veld foods (plants in open grasslands of South Africa) has its own choa ceremony (Thomas, 1959). The !Kung observe a first-fruit ceremony dealing with tubers. The rite is performed by the head man on a selected day. One of the prayers translates: "Father, I come to you, I pray to you, please give me food and all things that I may live" (Schapera, 1951). The tubers must not be touched until the ceremony is performed.

Spencer (1928, 1967) describes, in some detail, yam ceremonies on Melville Island, Australia. These are celebrated as rainy season initiation rites. One particular yam, called Kolamma or Kulemma, has small rootlets (like whiskers) all over it. It is supposed to make whiskers grow on boys and so is involved in growing-up rites. Girls may be initiated at the same time, but no female can touch the yam or the ceremonial fire until the rites are completed. One of the lines chanted is: "Yams, you are our fathers!" The natives assert that after the ceremony all kinds of yams will grow plentifully.

It might be mentioned here that the New Yam ceremonies are the most important in the ceremonial calendar of yam-eating tribes of West Africa. It is important not to dig some species of *Dioscorea* too early in the season and this sound agricultural practice is reinforced by religious ritual. A similar protective ritual is observed by the nonagricultural Andamanese (Coursey, 1972).

The Warramunga tribe of Australia has a yam totem; the Kaitisha tribe has a grass seed totem and celebrates a grass seed dance and ceremony. Rain dances are performed by both Bushmen and Australian natives to increase food resources. These are but a few of the many examples that could be given to show how plants that are important sources of food or well-being are venerated and intimately woven into the religious and ritual life of gathering peoples.

On Sharing the Bounty

Much has been made by some social anthropologists of the more or less egalitarian nature of hunter-gatherer societies. Game brought in by a hunter is shared by usually strict rules; the hunter himself has little to say about it. Produce gathered by the women and children is normally shared by the whole camp. One may own personal items like digging sticks, bows, arrows, spears, boomerangs, carrying nets, bowls, pots, and so forth, but if someone asks for something, it is very difficult to refuse. There is, in many tribes, a sense of community ownership, but this is likely to vary by degrees of relationship. Nearly all items might be shared by the immediate family or even within an extended family, but sharing becomes diluted with more distant relations. Lee (1988) pointed out that there are strong peer pressures to share freely and to prevent an individual from dominating or even showing pride. Egos are continuously being deflated by bawdy jokes and condescending comments.

The sharing rules may, in fact, cause the fragmentation of large camps. When a game animal is divided into so many shares that each person receives only a tiny scrap of meat, the better hunters are likely to fission off with their immediate families in order not to share with all. People may also resort to hiding personal items in order not to share. The social ideal of sharing freely has appeal, but seems to work best in small intimate groups.

Unharvested resources have a different set of rules. In many areas tribal territories are clearly defined, and even foraging microbands or families may be allotted specified regions, groves, or stands of useful plants. They very rarely harvest on land reserved for other bands. Springs and water holes may be owned by specified groups and the outsider must ask permission even to drink. This is true in both South Africa and Australia. In Australia, land ownership was respected during the burning season. It was considered a serious offense to burn another's foraging range. The time and place of a burn were carefully chosen and serious attempts were made to keep the fires within prescribed bounds (Warner, 1958).

If a pygmy finds a bee tree, he can mark it, and he alone is entitled to harvest the honey. To steal from a marked tree is a serious offense. Among Bushmen the same holds true for ostrich nests as well as bee trees and stealing either one can be punished by death. Some of the hollow trees of Bushmen ranges fill with water and provide an important source of water in a semidesert land. The trees may be individually owned and inheritance may pass from father to son (Marshall, 1960).

Tree marking is also observed in Australia (Gregory, 1886):

> A native discovering a *Zamia* fruit unripe will put his mark on it and no other native will touch this; the original finder of the fruit may rest perfectly certain that when it becomes ripe he has only to go and fetch it for himself.

Property rights are demonstrated by the custom of breaking off the top of the "grass tree" (*Xanthorrea*), which will then rear large edible grubs. The one who breaks off the top owns the grubs that will be produced later (Grey, 1841).

At least one Aboriginal family is reported to have owned a rock quarry. The head of the family removed slabs of rock, broke them into appropriate pieces, and shaped them crudely as blanks from which ax-heads could be made. The blanks and spears entered into the long-distance trading routes established by the Aborigines long before European contact (Coon, 1971).

A number of tribes of the Pacific Northwest kept slaves. These were captured in raids on neighboring tribes, purchased, received as gifts in pot-latch celebrations, or sometimes generated by voluntary servitude to settle debts. During the extravagant pot-latches of the 19th century, slaves were sometimes killed as a show of wealth. These rather sedentary tribes had a surplus of goods and commodities which were either distributed or simply destroyed. Such a luxuriant economy does not fit the stereotype of the starving savage (Suttles, 1968).

Also in the same region, houses, not necessarily made of skin or bark, were often individually owned, and some tribes built solid plank houses that were intended to be permanent structures. The Modoc (California–Oregon border) maintained a scheduled round of nomadic movements to exploit various resources at different times of the year. In winter camp, they lived in plank houses that were dismantled and carefully stacked each spring when they moved to summer quarters. The houses were reassembled on their return in late fall or early winter (Ray, 1963).

It would appear that private ownership of resources was well understood by nonagricultural people and probably by preagricultural people as well. The concept of ownership was, and still is, widespread and deeply ingrained in many gathering societies.

Population Control and the Aged

As previously indicated, the evidence seems to show that populations of hunter-gatherers are maintained well below the carrying capacity of the range. This is, in part, what keeps the system so stable and durable. When crops fail, farmers die of starvation, but famine is not recorded among gatherers except where there has been a drastic disturbance by outside agents (Coon, 1971):

> In every well-documented instance, cases of hardship may be traced to the intervention of modern intruders. Starvation came to the

Caribou Eskimo only after a few Cree Indians, armed with automatic rifles, had slaughtered a whole migration of caribou to cut out their tongues to sell to white canners.

What methods are used to keep the population stable? There seems to be little consistency in methodology; the only generality seems to be that some method or combination of methods is employed by each group. Infanticide is common, but far from universal. Since males are usually preferred to females, the practice may result in markedly displaced sex ratios in the population. Invalidicide is widespread, although some tribes treat the sick and injured with consideration and do not withhold customary medicines. Delayed marriage, late weaning, and wide spacing of children are among the most common methods of population control, and computer studies have shown that these alone can adequately stabilize a population (Skolnick & Cannings, 1972). Geronticide (killing of the aged) is also practiced in some tribes. In addition, warfare, raids, feuds, and similar activities often affect population size.

In general, there seems to be no model that has very wide application. Lee (1968) specifically investigated the situation of the aged among the Bushmen: "In a total population of 466, no fewer than 46 individuals (17 men and 29 women) were determined to be over 60 years of age, a proportion that compares favorably to the percentage of elderly in industrialized populations."

It is evident, then, that the "nasty, brutish, and short" stereotype of the hunting–gathering life styles was a product of an egocentric sense of superiority and that all features of it are demolished by serious anthropological studies.

Conclusions

The ethnographic evidence indicates that people who do not farm do about everything that farmers do, but they do not work as hard. Gatherers clear or alter vegetation with fire, sow seeds, plant tubers, protect plants, own tracts of land, houses, slaves, or individual trees, celebrate first-fruit ceremonies, pray for rain, and petition for increased yield and abundant harvest. They spin fibers, weave cloth, and make string, cord, baskets, canoes, shields, spears, bows and arrows, and ritual objects, recite poetry, play musical instruments, sing, chant, perform dances, and memorize legends. They harvest grass seeds, thresh, winnow, and grind them into flour. They do the same with seeds of legumes, chenopods, cucurbits, crucifers, composites, and palms. They dig roots and tubers. They detoxify poisonous plants for food and extract poisons to stun fish or kill game. They are familiar with a variety of drugs and medicinal plants. They understand the life cycles of plants, know the seasons of the year, and when and where the natural plant food resources can be harvested in greatest abundance with the least effort.

There is evidence that the diet of gathering peoples was better than that of cultivators, that starvation was rare, that their health status was generally superior, that there was a lower incidence of chronic disease (Lee & De-Vore, 1968), and not nearly as many cavities in their teeth (Angel, 1984).

The question must be raised: Why farm? Why give up the 20-hr work week and the fun of hunting to toil in the sun? Why work harder for food less nutritious and a supply more capricious? Why invite famine, plague, pestilence, and crowded living conditions? Why abandon the Golden Age and take up the burden?

References

Akazawa, T., & Aikens, C. M. (1986). *Prehistoric hunter-gatherers in Japan.* Tokyo, Japan: University of Tokyo Press.

Angel, J. L. (1984). Health as crucial factor in the changes from hunting to developed farming in the eastern Mediterranean. In M. N. Cohen & G. J. Armelagos (Eds.), *Paleopathology at the origins of agriculture* (pp. 51–73). New York, NY: Academic Press.

Bancroft, J. (1884). Food of the Aboriginees of central Australia. *Proceedings of the Royal Society of Queensland, 1*, 104–106.

Bicchieri, M. (1972). *Hunters and gatherers today.* New York, NY: Holt, Rinehart, and Winston.

Burkill, I. H. (1935). *A dictionary of the economic products of the Malay Peninsula.* London: Oxford University Press.

Campbell, A. H. (1965). Elementary food production by the Australian Aboriginees. *Mankind, 6*, 206–211.

Chase, A. K. (1989). Domestication and domiculture in northern Australia: A social perspective. In D. R. Harris & G. C. Hillman (Eds.), *Foraging and farming: The evolution of plant exploitation* (pp. 42–54). London: Unwin Hyman.

Coon, C. S. (1971). *The hunting peoples.* Boston, MA: Little, Brown, and Co.

Coursey, D. G. (1972). The civilizations of the yam: Interrelationships of man and yams in Africa and the Indo-Pacific region. *Archaeology and Physical Anthropology of Oceania, 7*, 215–233.

Cribb, A. B., & Cribb, J. W. (1975). *Wild food in Australia*. Sydney, Australia: Collins.

Dahlberg, F. (1981). *Woman the gatherer*. New Haven, CT: Yale University Press.

Downs, J. (1964). Significance of environmental manipulation in Great Basin cultural development. In W. L. D'Azvedo, W. A. Davis, D. D. Fowler, & W. Suttles (Eds.), *The current status of anthropological research in the Great Basin* (pp. 39–56). Reno, NV: Desert Research Institute.

Drucker, P. (1963). *Indians of the Northwest Coast*. Garden City, NY: Natural History Press. https://doi.org/10.5962/bhl.title.68277

Fernald, M. L., & Kinsey, A. C. (1943). *Edible wild plants of eastern North America*. Cornwall-on-Hudson, NY: Idlewild Press.

Flannery, K. V. (1968). Archaeological systems theory and early Mesoamerica. In B. J. Meggers (Ed.), *Anthropological archaeology in the Americas* (pp. 67–87). Washington, DC: Anthropology Society.

Foley, R. (1988). Homonids, humans and hunter-gatherers: An evolutionary perspective. In T. Ingold, D. Riches, & J. Woodburn (Eds.), *Hunters and gatherers I: History, evolution and social change* (pp. 207–221). New York, NY: Berg Oxford.

Fox, F. W., & Young, M. E. N. (1982). *Food from the veld: Edible wild plants of southern Africa botanically identified and described*. Johannesberg, South Africa: Delta Books.

Gentry, H. S. (1942). Rio Mayo plants: A study of the flora and vegetation of the valley of Rio Mayo, Sonora. *Botanical Gazette, 104*(4), 652.

Gifford, E. W. (1967). *Ethnographic notes on the southwestern Pomo* (pp. 1–48). Berkeley, CA: University of California Press.

Gregory, A. C. (1886). Memoranda on the Aboriginees of Australia. *Journal of Anthropological Institute, 16*, 131–133.

Grey, G. (1841). *Journals of two expeditions of discovery in northwest and western Australia during the years 1837, 38 and 39 under authority of Her Majesty's government* (Vol. 2). London: T. & W. Boone.

Hallam, S. J. (1975). *Fire and hearth: A study of Aboriginal usage and European usurpation in southwestern Australia*. Canberra, Australia: Australian Institute of Aboriginal Studies.

Harlan, J. R. (1967). A wild wheat harvest in Turkey. *Archaeology, 20*, 197–201.

Harlan, J. R. (1989). Wild grass-seed harvesting in the Sahara and sub-Sahara of Africa. In D. R. Harris & G. C. Hillman (Eds.), *Foraging and farming: The evolution of plant exploitation* (pp. 79–90). London: Unwin Hyman.

Harrington, J. P. (1932). *Tobacco among the Karuk Indians of California. Bull. 284*. Washington, DC: Smithsonian Institute Bureau of Ethnology.

Harris, D. R., & Hillman, G. C. (1989). *Foraging and farming: The evolution of plant exploitation*. London: Unwin Hyman.

Hayden, B. (1981). Subsistence and ecological adaptations of modern hunter/ gatherers. In R. Harding & G. Teleki (Eds.), *Omnivorous primates* (pp. 344–421). New York, NY: Columbia University Press. https://doi. org/10.7312/hard92188-011

Hesiod (1983). *The poems of Hesiod* (R. M. Frazer, trans.). Norman, OK: University of Oklahoma Press.

Ho, P.-T. (1969). The loess and the origin of Chinese agriculture. *The American Historical Review, 75*(1), 1–36. https://doi.org/10.2307/1841914

Hublin, J.-J., Ben-Ncer, A., Bailey, S. E., Freidline, S. E., Neubauer, S., Skinner, M. M., . . . Gunz, P. (2017a). New fossils from Jebel Irhoud, Morocco and the pan-African origin of *Homo sapiens. Nature, 546,* 289–292. https://doi. org/10.1038/nature22336

Hublin, J.-J., Ben-Ncer, A., Bailey, S. E., Freidline, S. E., Neubauer, S., Skinner, M. M., . . . Gunz, P. (2017b). Author correction: New fossils from Jebel Irhoud, Morocco and the pan-African origin of *Homo sapiens. Nature, 558,* E6. https://doi.org/10.1038/s41586-018-0166-3

Ingold, T., Riches, D., & Woodburn, J. (1988a). *Hunters and gatherers I: History, evolution and social change.* New York, NY: Berg Oxford.

Ingold, T., Riches, D., & Woodburn, J. (1988b). *Hunters and gatherers II: Property, power and ideology.* New York, NY: Berg Oxford.

Irvine, F. R. (1957). Wild and emergency foods of Australian and Tasmanian Aborigines. *Oceania, 28*(2), 113–142. https://doi.org/10.1002/j.1834-4461.1957. tb00733.x

Jardin, C. (1967). *List of foods used in Africa.* Rome, Italy: FAO.

Johnson, A. W., & Earle, T. (1987). *The evolution of human societies: From foraging group to agrarian state.* Stanford, CA: Stanford University Press.

Johnston, T., & Cleland, J. B. (1933). The history of the Aboriginal narcotic, pituri. *Oceania, 4,* 201–223. 268–289. doi:https://doi.org/10.1002/j.1834-4461.1933. tb00101.x

Jones, R. (1969). *Firestick farming.* Sydney, Australia: Australian Natural History Museum.

Klimek, S. (1935). *Culture element distributions. I. The structure of California Indian culture.* Berkeley, CA: University of California Press.

Körnicke, F. (1985). *Die Arten und Varietäten des Getreides.* Handbuch des Getreideaues, Vol. 1, Paul Parey, Berlin.

Koyama, S., & Thomas, D. H. (1982). *Affluent foragers: Pacific coasts east and west.* Osaka, Japan: National Museum of Ethnology.

Lee, R. B. (1968). What hunters do for a living, or how to make out on scarce resources. In R. B. Lee & I. DeVore (Eds.), *Man the hunter* (pp. 30–48). Chicago, IL: Aldine.

Lee, R. B. (1984). *The Dobe!Kung.* New York, NY: Holt, Rinehart and Winston.

Lee, R. B. (1988). Reflections on primitive communism. In T. Ingold, D. Riches, & J. Woodburn (Eds.), *Hunters and gatherers II: Property, power and ideology* (pp. 252–268). New York, NY: Berg Oxford.

Lee, R. B., & DeVore, I. (1968). *Man the hunter.* Chicago, IL: Aldine.

Lévi-Strauss, C. (1950). The use of wild plants in tropical South America. *Economic Botany, 6,* 465–486.

Levitt, D. (1981). *Plants and people: Aboriginal uses of plants on Groote Eylandt.* Canberra, Australia: Australian Institute of Aboriginal Studies.

Lewis, H. T. (1989). Ecological and technological knowledge of fire: Aborigines versus park rangers in northern Australia. *American Anthropologist, 91,* 940–961. https://doi.org/10.1525/aa.1989.91.4.02a00080

Loeb, E. M. (1934). *The eastern Kuksu cult. University of California Publications in American Archaeology and Ethnology.* (Vol. *33*(2), pp. 139–231). Berkeley, CA: University of California Publications.

Lourandos, H. (1980). Changes or stability? Hydraulics, hunter-gatherers, and population in temperate Australia. *World Archaeology, 11,* 245–264. https://doi.org/10.1080/00438243.1980.9979765

Maiden, J. H. (1889). *Useful native plants of Australia.* London: Trübner and Co. https://doi.org/10.5962/bhl.title.120959

Marshall, L. (1960). Kung Bushman bands. *Africa (London), 30*(4), 325–355. https://doi.org/10.2307/1157596

Murdock, G. P. (1967). The ethnographic atlas: A summary. *Ethnology, 6*(2), 109–236. https://doi.org/10.2307/3772751

Nabhan, G., & deWet, J. M. J. (1984). *Panicum sonorum* in Sonoran desert agriculture. *Economic Botany, 38,* 65–82. https://doi.org/10.1007/BF02904417

Palmer, E. (1883). On the plants used by the natives of North Queensland, Flinders and Mitchell Rivers for food and medicine, etc. *Journal and Proceedings of the Royal Society of New South Wales, 17,* 93–113.

Price, T. D., & Brown, J. (1985). *Prehistoric hunter-gatherers: The emergence of social and cultural complexity.* New York, NY: Academic Press.

Ray, V. F. (1963). *Primitive pragmatists: The Modoc Indians of Northern California.* Seattle, WA: University of Washington Press.

Riches, D. (1982). *Northern nomadic hunter-gatherers: A humanistic approach.* London: Academic Press.

Roy, S. C. (1921). A preliminary classification of wild rices of Central Provinces and Bihar. *Agricultural Journal of India, 16,* 365–380.

Sahlins, M. (1968). Notes on the original affluent society. In R. B. Lee & I. DeVore (Eds.), *Man the hunter* (pp. 85–89). Chicago, IL: Aldine.

Schapera, I. (1951). *The Khoisan peoples of South Africa.* London: Routledge and Kegan Paul Ltd.

Scheinsohn, V. (2003). Hunter-gatherer archaeology in South America. *Annual Review of Anthropology, 32*, 339–361. https://doi.org/10.1146/annurev. anthro.32.061002.093228

Schire, C. (1984). *Past and present in hunter-gatherers studies.* London: Academic Press.

Silberbauer, G. B. (1981). *Hunter and habitat in the central Kalahari Desert.* Cambridge, UK: Cambridge University Press.

Skolnick, M. H., & Cannings, C. (1972). Natural regulation of numbers in primitive human populations. *Nature, 239*, 287–288. https://doi.org/10.1038/239287a0

Spencer, B. (1967). *Wanderings in wild Australia* (Vol. *2*). London: Johnson Reprint Co.

Spencer, B. W. (1928). *Wanderings in wild Australia* (Vol. *1*). London: Macmillan.

Steward, J. H. (1934). *Ethnography of the Owens Valley Paiute. Publications in American Archaeology and Ethnology, 33*, 233–250.

Steward, J. H. (1941). *Culture element distributions: XIII Nevada Shoshoni.* Berkeley, CA: University of California Press.

Stickney, G. P. (1896). Indian uses of wild rice. *American Anthropologist, 9*(4), 115–122. https://doi.org/10.1525/aa.1896.9.4.02a00000

Story, D. (1958). *Some plants used by the Bushmen in obtaining food and water.* Pretoria, South Africa: South African Government.

Suttles, W. (1968). Coping with abundance: Subsistence on the Northwest Coast. In R. B. Lee & I. De Vore (Eds.), *Man the hunter* (pp. 56–68). Chicago, IL: Aldine.

Szafer, W. (1966). Foundations of a geobotanical division of Poland. In *The vegetation of Poland* (pp. 640–653). Oxford: Pergamon.

Thomas, E. M. (1959). *The harmless people.* New York, NY: Knopf.

Walters, I. (1989). Intensified fishery production at Moreton Bay, southeast Queensland, in the late Holocene. *Antiquity, 63*(239), 215–224. https://doi. org/10.1017/S0003598X00075943

Warner, W. L. (1958). *A black civilization: A social study of an Australian tribe.* New York, NY: Harper and Row.

Williams, N., & Hunn, E. (1982). *Resource managers: North American and Australian hunter-gatherers.* Boulder, CO: Westview Press.

Winterhalder, B., & Smith, E. A. (1981). *Hunter-gatherer foraging strategies: Ethnographic and archeological analyses.* Chicago, IL: University of Chicago Press.

Woodburn, J. (1970). *Hunters and gatherers: The material culture of the nomadic Hadza.* London: British Museum.

Woodburn, J. (1988). African hunter-gatherer social organization: Is it best understood as a product of encapsulation? In T. Ingold, D. Riches, & J. Woodburn (Eds.), *Hunters and gatherers II: Property, power and ideology* (pp. 31–64). New York, NY: Berg Oxford.

Yanovsky, E. (1936). *Food plants of the North American Indians. USDA Miscellaneous Publications No 237.* Washington, DC: US Government Printing Office.

Yen, D. E. (1989). The domestication of environment. In D. R. Harris & G. C. Hillman (Eds.), *Foraging and farming: The evolution of plant exploitation* (pp. 55–75). London: Unwin Hyman.

2

Views on Agricultural Origins

Men ought not presently to believe all they hear, but neither should they be as incredulous as I have sometimes been.

Friar Domingo Navarrate, 1676 (Cummins, 1962)

Agriculture as Divine Gift

In the classical mythologies of all civilizations, agriculture is fundamentally of divine origin. It arrived in different ways from different deities and under various circumstances, but the underlying theme is recognizable. In the Mediterranean region, the source was a goddess: Isis in Egypt, Demeter in Greece, and Ceres in Rome. In China, it was the ox-headed god Shên-nung; in Mexico, Quetzalcoatl disguised as a plumed serpent or other animal. In Peru, Viracocha or perhaps the Inca sent by his Father the Sun, was

Harlan's Crops and Man: People, Plants and Their Domestication, Third Edition.
H. Thomas Stalker, Marilyn L. Warburton, and Jack R. Harlan.
© 2021 American Society of Agronomy, Inc. and Crop Science Society of America, Inc.
Published 2021 by John Wiley & Sons, Inc.
doi:10.2135/harlancrops

responsible. The appearance of agriculture in mythology was almost always associated with other features of civilization: settled life, household arts, formal religion, and government by laws. We shall also see that agriculture brought death and gods that demanded sacrifice in exchange for rain and abundant harvests. The general features of these stories can be grasped from the selections that follow.

According to Diodorus Siculus (3000 years ago) agriculture originated in this way:

> Five gods were born to Jupiter and Juno, among them Osiris and Isis. Osiris married his sister Isis and did many things of service to the social life of man. Osiris was the first, they record, to make mankind give up cannibalism; for after Isis had discovered the fruit of both wheat and barley, which grew wild all over the land along with other plants, Osiris had also devised the cultivation of these fruits; all men were glad to change their food, both because of the pleasing nature of the newly-discovered grains and because it seemed to their advantage to refrain from their butchery of one another. As proof of the discovery of these fruits they offer the following ancient custom that they still observe; even yet at harvest time the people make a dedication to the first heads of the grain to be cut, and standing beside the sheaf beat themselves and call upon Isis, by this act rendering honor to the goddess for the fruits which she discovered, at the season when she first did this. Moreover, in some cities, during the festival of Isis as well, stalks of wheat and barley are carried among the other objects in the procession, as a memorial of what the goddess so ingeniously discovered at the beginning. Isis also established laws, they say, in accordance with which the people regularly dispense justice to one another and are led to refrain through fear of punishment from illegal violence and insolence; and it is for this reason also that the early Greeks gave Demeter the name Thesmophorus, that is lawgiver, acknowledging in this way that she had first established their laws.

C.H. Oldfather translation (1946)

It was Demeter who taught Tritolemous "to yoke oxen and to till the soil and gave him the first grains to sow. In the rich plains about Eleusis he reaped the first harvest of grain ever grown, and there, too, he built the earliest threshing floor. . . . In a cart given him by Demeter and drawn by winged dragons he flew from land to land scattering seed for the use of men" (Fox, 1916).

Half a world away, we find a myth containing exactly the same elements: (a) people without agriculture are savages who live like animals and eat each other; (b) through some divine instruction they not only learn how to produce food, but also to live by laws and to practice religion and those household arts common to civilized life. From the Royal Commentaries of the Inca Garcilaso de la Vega (1961) we read:

> Know then that, at one time, all the land you see about you was nothing but mountains and desolate cliffs. The people lived like wild beasts, with neither order nor religion, neither villages nor houses, neither fields nor clothing, for they had no knowledge of either wool or cotton. Brought together haphazardly in groups of two or three, they lived in grottoes and caves and like wild game, fed upon grass and roots, wild fruits, and even human flesh. They covered their nakedness with the bark and leaves of trees, or with the skins of animals. Some even went unclothed. And as for women, they possessed none who were recognized as their very own.
>
> Seeing the condition they were in, our father the Sun was ashamed for them, and he decided to send one of his sons and one of his daughters from heaven to earth, in order that they might teach men to adore him and acknowledge him as their god; to obey his laws and precepts as every reasonable creature must do; to build houses and assemble together in villages; to till the soil, sow the seed, raise animals, and enjoy the fruits of their labors like human beings.

The Inca king and queen arrived from heaven and were given a sign by which they would know where to establish a capital city. The place was located (Cuzco) and they set out to teach the savages "how to live, how to clothe and feed themselves like men, instead of like animals." The epic continues (from Garcilaso de la Vega, 1961 edition):

> While peopling the city, our Inca taught the male Indians the tasks that were to be theirs, such as selecting seeds and tilling the soil. He taught them how to make hoes, how to irrigate their fields by means of canals that connected natural streams, and even to make these same shoes that we wear today. The queen, meanwhile, was teaching the women how to spin and weave wool and cotton, how to make clothing as well as other domestic tasks.
>
> In short, our sovereigns, the Inca king, who was master of men, and Queen Coya, who was mistress of the women taught their subjects everything that had to do with human living.

The basic theme is repeated with regularity around the world. From cuneiform tablets, we learn that the source of agriculture for the Babylonians, Chaldeans, and Phoenicians was a god named Oannes who appeared to the inhabitants of the Persian Gulf Coast and instructed them on growing crops and raising animals (Fiore, 1965). According to Maurice (1795):

> He also taught men to associate in cities, and to erect temples to the gods, he initiated them in the principles of legislation, and the elements of geometry. He showed them how to practice botany and husbandry; and he reformed and civilized the first rude and barbarous race of mortals.

In Chinese mythology, P'an Ku separated the heavens and the earth, created the sun, moon, and stars, and produced the plants and animals. There followed 12 (or 13) celestial sovereigns, all brothers, who ruled 18,000 years each, then 11 terrestrial sovereigns, all brothers, who ruled 18,000 years each. After that came nine human rulers, all brothers, who governed a total of 45,600 years. Among these was Shên-nung, who taught the people agriculture and developed medicine. In another version, 16 rulers came after the nine and these were then followed by the "Tree Sovereigns," one of whom was Shên-nung. There are many variations of this particular theme (Christie, 1983; Fitzgerald, 1950; Latourette, 1941), including the following description of Shên-nung by the ancient Chinese historian Se-me-Tsien (3000 years ago). Shên-nung, he said, had the body of a man and the head of an ox, and his element was fire. He taught the people to use the hoe and the plow and initiated the sacrifice at the end of the year. He also found drug plants that cured and made a five-stringed lute (Chavannes, 1967).

In later Chinese history, Shên-nung is considered to have been an emperor, and a fictitious date (usually about 4800 BP) was assigned to his reign. He is said to have instituted the custom of ritually sowing five kinds of grains at the time of spring planting. The custom was preserved as late as the 20th century and the emperor himself participated in the ceremony. Actually, there is no evidence that there ever was a ruler by that name and the date is far earlier than any real date recorded in Chinese history.

The ancient legends have been amplified over the centuries and a veneer of embellishment has been added to the classical myths. The date given to Shên-nung is nonsense, but the myth of divine origin of agriculture is typical.

The mythologies of the Native Americans are enormously varied and complex, but here I shall only present themes of the Aztec and Maya to compare with the Incan myth already cited. In the Aztec creation literature, Quetzalcoatl was described as (from Prescott, 1936):

God of the air, a divinity who, during his residence on earth, instructed the natives in the use of metals, in agriculture, and in the arts of government. . . . Under him, the earth teemed with fruits and flowers, without the pains of culture. An ear of Indian corn was as much as a single man could carry. The cotton, as it grew, took, of its own accord, the rich dyes of human art. The air was filled with intoxicating perfumes and the sweet melody of birds. In short, these were the halcyon days, which find a place in the mythic systems of so many nations in the Old World. It was the *golden age* of Anahuac.

Interestingly enough, both the Aztec and the Maya thought that maize (*Zea mays* L.) was on earth before mortal men. In the Aztecan story, Quetzalcoatl disguised himself as a black ant, stole the cereal from Tonacatepel, and took it to Tamoanchan for the benefit of man. In the Mayan creation myth, the flesh of man was actually formed out of maize meal and snake's blood (Recinos, 1947). It is little wonder that the maize plant is venerated to this day in Mexico and Guatemala. The Mayan epic also contains oblique references to a garden of Eden or golden age in which nature yielded abundantly of its own accord.

> In this manner they were filled with pleasure because they had discovered a lovely land full of delights, abundant in yellow ears and white ears (of maize) and also abundant in (two kinds of) cacao and innumerable fruits of mamey, chirimoya, jocote, nance, white zopote, and honey. The foods of Paxil y Cayalá were abundant and delicious.
>
> *Popol Vuh pt. III,* as reported in Recinos (1947)
> (J. R. Harlan translation)

Note that the fruits in the Popal Vah text were believed to be *Lucuma mammosa* C. F. Gaertn., *Annona cherimola* Mill., *Spondias purpurea* L., *Byrsonima crassifolia* (L.) Kunth, and *Casimiroa edulis* La Llave & Lex., respectively.

In all the myths and tales mentioned so far, and many like them, the knowledge of agriculture is gratefully received as a blessing from the gods. The outstanding exception is found in Genesis where agriculture comes as a curse:

> 3:17 . . . cursed is the ground for thy sake; in sorrow shalt thou eat of it all the days of thy life; 3:18 Thorns also and thistles shall it bring forth to thee; and thou shalt eat the herb of the field; 3:19 In the sweat of thy face shalt thou eat bread, till thou return unto the ground; for out of it wast thou taken: for dust thou art, and unto dust shalt thou return.

3:22 And the Lord God said, Behold, the man is become as one of us, to know good and evil: and now, lest he put forth his hand, and take also of the tree of life, and eat, and live for ever: 3:23 Therefore the Lord God sent him forth from the Garden of Eden, to till the ground from whence he was taken.

King James Version

There is no need to comment on all the various mythologies of agricultural peoples, but lest one be tempted to make too much of the similarities and underlying themes, I must point out that the Australian Aborigines, who did not practice agriculture, also had their mythologies and creation stories in which gods taught the people how to gather foods. An elderly Aborigine woman recited this part of the creation legend (as reported by Berndt & Berndt, 1970):

Ngalgulerg (a mythical woman) gave us women the digging stick and the basket we hang from our foreheads, and Gulubar Kangaroo gave men the spear-thrower. But that Snake that we call Gagag (Mother's mother)—taught us how to dig for food and how to eat it, good foods and bitter foods.

Except for Genesis, the stories of agriculture as divine gift support the stereotype described in the previous chapter. The consensus of agricultural people is that:

1) There was a time before agriculture when people gathered their food from the wild.
2) Not farming is primitive, wild, uncivilized, lawless, graceless, and brutish.
3) Nonfarmers did not farm because of ignorance of lack of intelligence.
4) A god or a goddess was required to enlighten them as to agricultural practices as well as laws, arts, religion, and civilized behavior.
5) Agricultural man knew himself to be superior to hunter-gatherers.

While the fallacy of all this is demonstrable, this way of thinking has persisted to the present time and has colored modern concepts of agricultural origins. For example, it has been argued that vegetatively reproduced crops *must* be older than seed crops because it would be easier to think of; it would not occur to the savage mind that seeds could be planted. Another product of this way of thinking is the idea that it could have happened only once or twice at most. If we can rid ourselves of the stereotype, more possibilities open up.

Table 2.1 Major time periods delineated by years before present (BP).

Paleolithic	2.6 million to 12,000-10,000 BP
Pleistocene	2.6 million to 11,700 BP
Mesolithic	11,600 BP (end of last ice age) to when agriculture starts (SE Europe: 9,000 BP; central Europe, 7500 BP; N Europe, 6000 BP; Near East, 11,000 BP)
Neolithic	11,000 BP to 8500 BP (Middle East), 3700 BP (N Europe)
Holocene	11,700 BP to present
Pre-pottery	10,500 to 7500 BP
Bronze	5300 to 3200 BP (2600 BP in Europe; 2300 BP in East Asia)
Iron	3200 BP to 2500 BP (started and ended later in European regions, lasted until 2300 BP in East Asia)
Ceramic	2800 to 2200 BP (Early — South America and Central America); 3500 to 2600 BP (Late — South and Central America)
Byzantine	3453 to 2330 BP
Classic	2900 to 2250 BP (Mesoamerica); 2500 to 3200 BP (South America)

Throughout the text there are references to various time periods in evolution and man's development and Table 2.1 can serve as a reference. Note that the same periods have different definitions or spans of time in different regions of the word.

Domestication for Religious Reasons

About 1900, Eduard Hahn (1896, 1909) proposed a theory that some animals might have been first domesticated out of religious concern rather than for economic reasons. He chose the urus (*Bos taurus* L.), a form of wild cattle, as his model, but the idea was extended to other animals and tentatively to plants (Anderson, 1954). The idea has not dominated anthropological thinking but has been revived from time to time and appears in anthropological and geographical literature. The possibilities are intriguing and the theory should be considered on its merits.

Hahn argued that it would have been impossible to predict the usefulness of domestic cattle before they were actually domesticated. Wild cattle are large and fierce beasts and no one could have foreseen their utility for labor or milk until they were tamed. What motivated man to take the initial steps? They were domesticated, argued Hahn, for ritual sacrifice in connection with lunar goddess cults, for the great curving horns of the urus were crescent shaped. We know that people from Western Europe to India have long held special religious feelings about cattle.

Even during the Ice Age, cattle were featured in the cave art of southwestern France and northeastern Spain. The great hall of the bulls at Lascaux is an eloquent testimony to the concern for wild cattle. The archaeological site of Catal Hüyük in Turkey, dating back into the seventh millennium BC, reveals a series of altars, one above the other, each featuring cattle heads. The animals are also depicted in painted murals on the temple walls. Much later, we find elegantly painted bull-vaulting scenes on the walls of temples at Knossos, Crete. Cattle were sacred to the Egyptians, were sacrificed by the Romans, and are still considered holy by the Hindus of India.

Indeed, to this day, we find a "bull belt" extending from Spain and Portugal to eastern India in which people have a special religious feeling about cattle. At the western end of this region, animals are publicly and ceremoniously slaughtered before thousands at the bullfight rituals, usually on Sundays. At the eastern end of the belt, naked Sadhus lead riots in favor of anti-slaughter laws that would protect cattle, and in the southern portion, deep into the Sahara and beyond, cattle-herding tribes have special, mystical attachments between man and beast.

Or, consider the mithan. This is another form of *Bos* (the taxonomy varies according to taxonomists) thought to have been domesticated from the wild gaur of India. Mithan are kept by hill tribes from Upper Myanmar westward across Assam, the Naga Hills, and into Bhutan. They are not herded, but allowed to range in the woods and meadows. They are, however, individually owned and fairly tame. They are not used for transportation, draft, or milk, but are raised for prestige, wealth, and sacrifice only. Mithan are used to purchase land and pay bride prices, fines, and ransoms. They are sacrificed at certain special religious observances, and sometimes, as a show of wealth, a rich man may sacrifice a number of animals in front of a rival's house to display wealth or humiliate an enemy. The animals are left where they are killed and others come and take away the remains to eat. Mithan are eaten, but only after ritual sacrifice. Skulls and horns are used to decorate temples, houses, and graves.

Chickens are thought to have been domesticated from the jungle fowl of southern and southeastern Asia. In parts of Asia, chickens are raised, but neither the flesh nor the eggs are eaten. The birds are used for sacrifice, divination by examining the entrails, or cock fighting. The art of divination from sacrificed birds seems to have spread with chicken raising at least into the Mediterranean area and was practiced by the ancient Greeks. The practice of rearing for sacrifice but not eating flesh or eggs has also been found in parts of the Americas and has led Carter (1971), Sauer (1952), and others to postulate early trans-Pacific contacts between the hemispheres.

Sheep, goats, pigs, and pigeons were sacrificed in the ancient world of classical times and it has been suggested that these also may have been domesticated to have a supply of sacrificial animals. From the above examples of legends and myths and from other clues, it seems at least plausible that animals may have been used in ritual killings as a substitute for humans. Human sacrifice and ritual killing may have been very ancient customs.

We know that there are a number of plants, wild and cultivated, that are used for ritual, ceremonial, and magical purposes. Some are drug plants, some produce dyes, and some have colorful leaves or flowers. I know of one plant of the West African forests which has a metallic, iridescent glint to the leaves and is used to mark the sites of secret (Poro) society meetings in the jungle. Anderson (1954) nominated the amaranths as candidates for ritual domestication. The blood-red inflorescences were used in religious ceremonies of ancient South America and I have seen them displayed over doorways in India and Pakistan. The pigment from another species is used in Hindu rituals. The Aztecs, among others used the grain in their rituals of human sacrifice, consuming popped seed in human blood.

Many narcotic and hallucinogenic plants have been used in religious ceremony and ritual. This, of course, does not mean that drug and ritual plants were domesticated *before* food plants, but it would not be wise in dealing with human affairs to ignore the motivations of religious concern.

Domestication by Crowding

Some decades ago, V. Gordon Childe proposed what came to be known as the "propinquity theory." Childe was a social-minded historian and prehistorian who was impressed by the evidence that the climates of North Africa and parts of the Near East had become increasingly desiccated over a period of several millennia BC. He visualized the rangelands drying up, forcing herd animals and man as well to withdraw to the banks of the few perennial rivers and to the oases where water could be found year-round. This brought man and animal into more intimate contact than had previously been the case and eventually induced man to domesticate some animal species (Childe, 1952).

In those days, many people still thought that man went through a set, three-phase development. He was first a hunter, then a herder, then a cultivator. The idea goes back to Greco-Roman times and still persists in some quarters. Having become a herder it was not difficult to pass to the next phase. The disturbance of the soil and vegetation by livestock at camp sites, together with manuring, would encourage weedy plants to grow. It was just such weeds that were said to be first taken into the domestic fold, and it was a short step from gathering them from the sheepfold to sowing them on purpose.

Childe (1925) also elaborated on what he called the "Neolithic revolution," that is, the shift from hunting and gathering to food production. He saw this as a radical and fundamental transformation of human adaptation and the most important development since the discovery of fire. The concept of an agricultural revolution has had more success than the oasis theory of domestication. The latter, however, was instrumental in stimulating a considerable amount of archaeological research because it was, to some degree, testable. Most of the testing was stimulated by the work of Robert J. Braidwood (1972) who set out to obtain archaeological evidence for the evolution of food production in the Near East. Many archaeologists have followed his example and there is now a large body of evidence on the subject. The evidence does not bear out the propinquity theory very well, but climate has changed and has altered the available food supplies and those changes must be taken into consideration.

Agriculture as Discovery

The most extensively developed model for agricultural origins is that cultivation was an invention or discovery. Because Darwin's theory of evolution has had profound influence on modern biology and anthropology, it is interesting to see how he viewed the subject (Darwin, 1896):

> The savage inhabitants of each land, having found out by many and hard trials what plants were useful, or could be rendered useful by various cooking processes, would after a time take the first step in cultivation by planting them near their usual abodes. . . . The next step in cultivation, and this would require but little forethought, would be to sow the seeds of useful plants; and as the soil near the hovels of natives would often be in some degree manured, improved varieties would sooner or later arise. Or a wild and unusually good variety of a native plant might attract the attention of some wise old savage; and he would transplant it, or sow its seed.

Darwin, among others, was convinced that nomadic people could not develop agriculture (Darwin, 1909):

> Nomadic habits, whether over wide plains, or through the dense forests of the tropics, or along the shores of the sea, have in every case

been highly detrimental (to "progress"). Whilst observing the barbarous inhabitants of Tierra del Fuego, it struck me that the possession of some property, a fixed abode, and the union of many families under a chief, were the indispensable requisites for civilization. Such habits almost necessitate the cultivation of the ground; and the first steps in cultivation would probably result, as I have shewn elsewhere (above), from some such accident as the seeds of a fruit tree falling on a heap of refuse, and producing an unusually fine variety.

Darwin (1909) concluded, however, that "the problem . . . of the first advance of savages towards civilization is at present much too difficult to be solved."

Elaborations on the theme developed the "happy accident" or *"Eureka!"* theory of plant domestication. No motive is required, only the brilliant revelation that seeds can be sown to produce plants when and where desired. The advantages of producing food on purpose are so obvious that all that was needed was the concept and then the development of agriculture was assured.

There are several ideas in the Darwinian view that should be separated for clarity: (a) man must be sedentary before he can cultivate plants; (b) useful plants are most likely to be discovered in manured refuse heaps; (c) useful plants are likely to be first planted in dump heaps; and (d) a wise old savage is required to start the process.

These concepts seem reasonable enough and have provided the basis for several theoretical treatments of the subject. One of the most influential was that of Carl O. Sauer (1952), a geographer whose *Agricultural Origins and Dispersals* has become a classic. He combined the Darwinian views with Eduard Hahn's idea (1896, 1909) that vegetative propagation should precede seed agriculture, and set out to locate the cradle of agriculture on theoretical grounds. He listed six presuppositions as a basis for his search (here condensed):

1) Agriculture did not originate from a growing or chronic shortage of food. People living in the shadow of famine do not have the means or time to undertake the slow and leisurely steps out of which a better and different food supply is to develop in a somewhat distant future.
2) The hearths of domestication are to be sought in areas of marked diversity of plants and animals. This implies well-diversified terrain and perhaps variety of climate.
3) Primitive cultivators could not establish themselves in large river valleys subject to lengthy floods and requiring protective dams, drainage, or irrigation.

4) Agriculture began in wooded lands. Primitive cultivators could readily open spaces for planting by deadening trees; they could not dig in sod or eradicate vigorous stoloniferous grasses.
5) The inventors of agriculture had previously acquired special skills in other directions that predisposed them to agricultural experiments.
6) Above all, the founders of agriculture were sedentary folk.

The sedentary life, he thought, could best be developed by fishing tribes, and for his purpose he sought them on fresh waters in a mild climate. Fresh water was selected because seaside vegetation has contributed relatively little to agriculture and what has been developed has come late in crop evolution. With these presuppositions in mind, he proposed Southeast Asia as the oldest hearth of agriculture. From there, systems spread northward into China and westward across India and the Near East, into Africa and the Mediterranean region, and finally into northern and western Europe. In the Americas, he located the original hearth in the northwestern part of South America from whence agriculture spread northward into Mexico, then to eastern North America, southward along the Andean chain, eastward to the Atlantic coast of Brazil, and to the Caribbean island chain. He left open the possibility that civilization might have been transmitted from the Old World to the New World.

Southeast Asia was selected because most anthropologists have felt that agriculture is older in Asia than in the Americas and because that region fits most of his presuppositions best. In particular, it had a mild climate and varied terrain, and was rich in fresh water aquatic resources as well as edible plants. People could settle down in permanent villages and develop the arts of cultivation without the pressures of periodic scarcity. The fact that a different set of plants was domesticated everywhere did not bother him. It was the *idea* of cultivation that diffused and that once people were shown the obvious superiority of the system, they would begin to domesticate plants from their own flora even if the rewards were to be found in the distant future.

Edgar Anderson (1954) liked Sauer's view and added some genetic threads to the fabric. He saw weeds as potential domesticates; he also thought that an increase in hybridization, with disturbed habitats, could result in increased variation and new genetic combinations from which useful selections could be made:

> Rivers are weed breeders; so is man, and many of the plants which follow us about have the look of belonging originally on gravel bars or mud-banks. If we now reconsider the kitchen middens of our sedentary fisherfolk, it seems that they would be a natural place where some of the aggressive plants from the riverbanks might find a home, where seeds and fruits brought back from up the hill or

down the river might sometimes sprout and to which even more rarely would be brought seeds from across the lake or from another island. Species which had never intermingled might do so there, and the open habitat of the rubbish-heap would be a more likely niche in which strange new mongrels could survive than any which had been there before man came along.

Anderson also felt that agriculture began in the tropics on dump heaps and that vegetative propagation predominated at the beginning, but he also left open the question of early transoceanic contact.

Evidence accumulated since the Sauer–Anderson models were suggested has indicated that some of their presuppositions were incorrect. For example, sedentary life is not essential to the evolution of agriculture. In Mesoamerica, there is good archaeological evidence that the people remained nomadic long after they were purposely growing plants for food (Flannery, 1968, 1986). In the Near East, there is evidence that a nuclear center developed in an area not in the tropics and by people not necessarily dependent upon aquatic resources. In that region, the people most dependent upon fishing and fowling, the Natufians, were among the last to take up agriculture. Thus, although the Sauer–Anderson models have been widely accepted by many, they are open to question.

Agriculture by Stress

A number of investigators have been persuaded that agriculture was adopted as a result of stress brought on by an increase in population and depletion of the foraging ranges. Mark N. Cohen is the most prominent of this school and he developed the argument at book length in his *The Food Crisis in Prehistory* (Cohen, 1977). He found archaeological evidence for depletion of local resources in the change of diet from preferred foods to those less preferred and less nutritious and in the exploitation of resources not used or little used in earlier times. These newer resources also may come from greater distances than from the sites excavated. The argument that present or recent hunter-gatherers keep the population well below the carrying capacity is countered by the argument that recent hunter-gatherers are not typical of preagricultural people. Those who could manage population size could remain hunter-gatherers; those who did not became farmers.

Later, Mark Cohen and George J. Armelagos organized a symposium on the paleopathology of people at the time when agriculture was being adopted in various places around the world (Cohen & Armelagos, 1984).

The reports in the symposium present some fascinating glimpses of the health status of ancient people. On the whole, they did not provide evidence for a decline in health before the adoption of agriculture, but there was a clear consensus that early farmers were not as healthy as preagricultural people. In general, the people of upper Paleolithic were taller, had excellent health, and no evidence of endemic disease.

In the eastern Mediterranean, there was a sudden drop in stature and evidence of some anemia and malaria during the Mesolithic. Presumably the rise in sea level resulted in more marshes and mosquitos and an increase in population density favored the virulent *Plasmodium falciparum*. The diet, however, appeared good (Angel, 1984).

The nutritional health of Neolithic people in that region was low and remained low for about 5000 years until a major improvement in classic times (2650–2003 BP). The decline was progressive, not sudden, and there was no evidence to suggest that man was forced into agriculture by a decline in diet. While irrelevant to agricultural origins, John L. Angel's data on teeth are instructive. Using number of lesions per mouth, that is, caries, abscess, and loss, he reported the following: Paleolithic: 2.3; Mesolithic: 1.3; Neolithic: 2.6–3.5; Bronze: 5.0–6.7; Iron: 6.8; Classic: 4.1; then fluctuating from 5.2 to 6.6 until the 19th century (except 3.4 in Byzantine times). In the 19th century, lesions per mouth jumped to 12.3 and the modern United States white population has nearly 16! Sugar has become cheap and abundant (Angel, 1984). We have, however, recovered the height we lost at the end of the Pleistocene, and a little more.

Agriculture as an Extension of Gathering

In Chapter 1, it became clear that hunter-gatherers have long known all they needed to know to develop agriculture. They did not need to discover the concepts of planting; they already had them. We have asked, "Why farm?" We could also ask the question, "Why not farm if you are equipped with all the materials and information to do so?" One approach is to ask a gatherer. During his study of the Bushmen, Richard Lee did exactly that, and he received the celebrated reply, "Why should I farm when there are so many mongongo nuts?" (Lee & DeVore, 1968). The Aborigines put it in almost the same terms (Berndt & Berndt, 1970):

> You people go to all that trouble, working and planting seeds, but we don't have to do that. All these things are there for us; the Ancestral Beings left them for us. In the end, you depend on the sun

and the rain just the same as we do, but the difference is that we just have to go and collect the food when it is ripe. We don't have all this other trouble.

Perhaps, even more to the point, an informant told Athol Kennedy Chase: "It is not our way; it is alright for other people. We get our food from the bush." (Chase, 1989). It is a question of what is perceived to be right and proper.

We now have some data to show that the Aboriginal opinion has merit. In 1965, Esther Boserup (1965) published a work entitled, *The Conditions of Agricultural Growth* that stimulated a number of studies on the input and output of energy in various systems. She showed that, overall, increasing energy inputs results in a decrease of output per amount of energy put into the system. David Pimentel and his coworkers at Cornell University followed with a long series of reports and a review that summarizes much of the work in this field (Pimentel & Hall, 1989). An important study by Black (1971) could also be cited. The most efficient agricultural systems use human labor only. For cassava in Zaire and Tonga, returns in kcal per kcal invested were 37.5 and 26.9, respectively. For sorghum in Sudan and maize in Mexico, returns were 14.1 and 10.1, respectively. Using draft animals returns were 3.3 for rice in Philippines, 3.4 for maize in Mexico, −0.5 for wheat in India, and −0.1 for sorghum in Nigeria. With high mechanization, figures for the United States are approximately 2.5 for maize, 1.4 for rice, 1.8 for wheat, and 2.3 for potato (Pimentel, 1974). Data from hunter-gatherers are confounded by different methods of calculation, but some results indicate returns comparable to or higher than the most efficient agricultural systems. My wild wheat harvest in Turkey yielded 40–50 kcal per kcal expended (Evans, 1975). The Biblical view of agriculture as a curse has support from these studies.

There is ample anthropological and ethnographical evidence to show that increasing the food supply through cultivation means an increase in work for the people supplying food. In general, the more intensive the agricultural system, the more work is required for a unit of food. Thus, if we are to understand the origins of agriculture, we must visualize situations in which man is willing to expend more energy to obtain food. In this respect, farming is not so attractive that gatherers are likely to take it up on sight or on first contact. Some rather compelling reasons would seem to be required.

In preagricultural times, the human population was not regulated by the food supply. If this were the case, then as Binford (1968) pointed out, there are two corollaries that would follow: "1) Man would be continually seeking means of increasing his food supply, and 2) It is only when man is freed from preoccupation with the food quest that he has time to elaborate culture." From what we have seen, both are patently false. Populations of hunter-gatherers are regulated well below the carrying capacity of the range, and the environment does not exert pressure on man to change his food procurement systems. Neither agricultural nor industrial man has anything like the leisure time of hunters and gatherers. Therefore, we must look elsewhere for the motivation to carry on agriculture.

What, then, might generate the motives that caused man to domesticate plants (and animals)? A much-cited model is one based on proposals put forth by Lewis Binford (1968) and Kent Flannery (1968). It attempts to integrate ethnographic and archaeological information and suggests not only reasons for but also places where the initiative toward food production might have occurred. Explicit in the Binford–Flannery model is the recognition that gatherers are sophisticated, applied botanists who know their materials and how to exploit them. They are prepared to grow plants if and when they think it would be worth the effort. Furthermore, the differences between intensive gathering and cultivation are minimal; recall the square kilometers of Australian landscape pitted by Aborigines digging yams.

Binford, in particular, emphasized the fact that one of the general post-Pleistocene adaptations of man was a fuller exploitation of aquatic resources. This is one of the most characteristic features of the so-called "Mesolithic" wherever it can be identified. Canoes, boats, and rafts were developed, and there was a great proliferation of archaeological sites that suggested permanent residence and subsistence by fishing, fowling, and gathering. The sedentary fisherfolk referred to by Sauer and Anderson did appear in many parts of the world; however, Binford suggests that it was not they who began domestication, but groups that budded off from them and who migrated into regions already occupied by hunter-gatherers. The argument goes that long before there was a food resource crisis among the fisherfolk, groups would move out and migrate into less well-endowed regions and ecological zones.

The fisherfolk population remained stable, but the migrants precipitated a crisis along the interface between the sedentary peoples and the nomadic hunter-gatherers. It was in response to this crisis that people were willing to go to the effort of cultivation.

The Binford model was spelled out in sufficient detail that he could make some predictions to be tested:

1) The initial activities of domestication in the Near East will appear adjacent to areas occupied by sedentary forager-fisherfolk (evidence for this was fairly firm at the time of the prediction).
2) Evidence of independent domestications will be found in European Russia and south-central Europe [good evidence of this has been reported, for example, Langlie et al. (2014) and Lisitsina (1984)].

3) Evidence of similar events will be found widely separated over Europe, Asia, and the Americas. [See Ford (1981) and Smith (1989) for reviews of early gardening in the Midwest USA]. Pickersgill (2007) among many others provided evidence from Mesoamerica and South America and the African evidence is compatible with the prediction.

There may be biological and ecological reasons as well for proposing that cultivation would begin adjacent to the best foraging ranges rather than in them. In the Near East, massive stands of wild wheats cover many square kilometers. Harlan and Zohary (1966) have asked, "Why should anyone cultivate a cereal where natural stands are as dense as a cultivated field? If wild cereal grasses can be harvested in unlimited quantities, why should anyone bother to till the soil and plant the seed?" The same arguments could well apply to the African savanna or to California, where wild food resources were abundant.

A major implication of the model is that the activities of plant domestication are likely to have taken place independently and probably simultaneously in many areas all over the world. The space–time pattern that would emerge would be almost the opposite of that of the Sauer–Anderson model. It would appear that the differences are testable by archaeological means and that even botanical and genetical evidence could come to bear on the problem.

The term "diffuse origins" (Harlan, 1956, 1961, 1980, 1986) is used for individual crops as well as to agricultural systems. Individual crops have origins that are diffuse in time and space in the sense that they evolve over time as they spread into new regions. At the beginning of domestication, they are like the wild forms, but the products may be enormously modified and found far from the original source or sources. Agriculture as a food-producing system is diffuse in the sense that we will not and cannot find a time or a place where it originated. We will not and cannot because it did not happen that way. Agriculture is not the result of a happening. It is not due to an idea, a discovery, an invention, a revelation, nor even a goddess. It is the end product of a long period of adaptive coevolution. The processes sometimes took millennia and were often spread over regions some thousands of kilometers across.

Domestication by Perception

A problem I have with the current theories about reasons for taking up farming is that they are all proposed by 20th century, university-educated, middle-class pragmatists all looking for some golden bottom line that would explain it all. Labor and time inputs, optimum foraging strategies, and so on are abstractions of the modern mind-set and world view. Could we come nearer to an understanding if we attempted to approach the mind-set and perceptions of the people who actually set the processes of domestication in motion? It seems to me that we might obtain some clues from perceptions of surviving hunter-gatherers and from folklore of subsistence farmers.

I have mentioned the perception of Aborigines that a landscape left unburnt for a number of years was, somehow, uncomfortable and inhabited by demons and malevolent spirits and was spiritually dangerous (Chase, 1989). Farmers in Amazonia have a similar perception of safe and dangerous space. To them, the forest is dangerous and full of demons and evil spirits; the house garden is safe and even protected by invisible harpy eagles (Reichel-Dolmatoff, 1971). The fields that produce most of the food for a village and migrate through the jungle year-by-year in a bush fallow rotation are perceived as intermediate in spiritual safety. The Kuruk and other tribes of western North America who grew tobacco and procured their food by hunting and gathering also had concepts of safe and danger-ous spaces. They were afraid of wild tobacco because it might have sprouted on the grave of someone and contain malevolent spirits. They grew their own ceremonial tobacco in a safe space. It is easy to see how such percep-tions would lead to gardening.

The perception of an association between plants like tobacco and the grave has a remarkable distribution. Consider the following folktales:

"A mother who lost her only daughter spent her days weeping at her grave. One day a strange plant sprouted from the grave. It grew taller and taller before her eyes. It was not good to eat after boiling it, roasting it or steaming it. . . . She tried smoking it and it comforted her". (Mayer, 1986, p. 278). This is the origin of tobacco in Japan.

In China, opium appeared on the grave of a wife who had been mistreated by her husband. When the husband was near death, she appeared to him in a dream and told him how to gather the latex and smoke it. He did as he was told and was comforted and cured of his illness—temporarily; if he did not smoke every day, he fell ill again to the point of near death (Eberhard, 1965). This explains addiction as well as origin of the crop.

In the Gran Chaco of South America, we are told: A cannibal woman is killed by a culture hero and from the ashes, the first tobacco grows (Wilbert, 1987, p. 151). A similar story is told about coca in South America and about the betel palm and the betel leaf in Southeast Asia and South Pacific.

Among the Fang of Gabon in Africa, *Tabernanthe iboga* L. is an important hallucinogen used in certain rituals and initiation ceremonies. The alkaloid, ibogaine, is extracted from the bark of the roots and is sufficiently potent that an occasional initiate is lost by overdose. It is said that a creator god killed a pygmy and cut off his fingers and toes, which he planted, and from the digits came this powerful plant (Dorson, 1972).

However, all these folktales about psychedelic plants belong to a larger family of stories concerning origins of food plants and of agriculture itself. Here is one from the Japanese Kojiki compiled in 712 AD (Mabuchi, 1964, p. 3).

A heavenly god asked an earthly goddess for a meal. Having seen her cooking various kinds of food taken out of her mouth, nose and anus, the heavenly god killed her in anger. Shortly afterward there appeared seeds of various crops from her corpse; from her eyes rice, from her ears the "millet," from her nose the red bean, from her anus the soya bean, from her vagina barley, while the silk worm came out of her head.

The tale is open-ended; for example, after maize was imported, it was added and was said to come from the teeth which are in rows like maize kernels. The source of soybean suggests flatulence, and other parts of the body are suggestive as well. In similar stories, the coconut comes from a human or monkey's head, bananas from fingers, and so on.

The following tale from New Guinea explains the origin of agriculture (Healey, 1988, p. 10):

A group of women lived alone in the grasslands. They had no gardens but ate game which they flushed from the grass by fire. One day a grass fire spread to the forest and burnt a menjawai forest demon in his lair in an epiphytic fern. After the fire had died down the women saw a column of smoke rising from the burnt forest. They went to investigate and found the smoking corpse of the demon. In fear they hurried back to their grasslands, but some months later they returned to find all manner of crops sprouting from the belly of the demon. They took and planted cuttings of the crops and experimented with various ways of preparing them before they discovered the proper ways to cook them.

There are hundreds of tales on the same theme with an essentially worldwide distribution. Someone or something must die for crops to appear and grow. In many tales death came for the first time with agriculture. As Mabuchi put it, "From the one who died the primordial death, there originated food plants, while human beings became mortal by this event. By repeating ritually such a primordial act, the fertility of both plants and human beings is to be secured. With this view are closely interrelated the human sacrifice, head-hunting, cannibalism, the ritual death in initiation ceremony and so on, death, killing, procreation and reproduction forming an inseparable unit." (Mabuchi, 1964, p. 85).

The crops of the Aztec were irrigated by human blood; thousands of victims were sacrificed yearly to appease the gods who controlled the weather and crop growth. The Phoenicians sacrificed their own children to Baal. This horrified the Hebrews who at some time substituted animals for humans, but the number of animals slaughtered is rather remarkable. From Leviticus 23 and Numbers 28–29 one can calculate a yearly requirement of 113 bulls, 37 rams, 1093 lambs, and 30 goats, by the priests alone. This does not include free will offerings, sin offerings, or guilt offerings volunteered by the people. By the time of Josephus, the number of rams required had increased to 118. The birth of agriculture was generally a bloody business.

The worldwide distribution of themes of origin tales tell us something of the perceptions and mind-set of the people who first took up the cultivation of plants and the taming of animals. These people lived in a world full of spirits, demons, and ogres. They did not view the world as we do and were not concerned with getting the most food for the least amount of effort or in the shortest possible time. Motivation was far more likely to have been in terms of what was perceived as spiritually safe and religiously comfortable. We do not know and never will know the perceptions of the Native Americans of Oaxaca who grew squashes (*Cucurbita* spp.) on a small scale in summer camp as they made their rounds as hunter-gatherers. Presumably, a wild squash is not so menacing as wild tobacco, but wild squash is very bitter. The bitterness may have been perceived as some kind of plant "power" worthy of respect and consideration. We shall never know what they thought, but it seems certain that a few squash vines had very little if any effect on the economy of the Native Americans who grew them. The activity could hardly have affected the food supply significantly. If they were being forced into agriculture because of population pressure, they surely would not have taken 2000 years to accomplish the job.

A No-Model Model

Every model proposed so far for agricultural origins or plant domestication has generated evidence against it. It is possible that some plants and animals were domesticated for ritual, magic, ceremony, or religious sacrifice, but only a few out of hundreds of species could be so identified. It is likely that a few cultigens did originate from dump heap weeds, but many show no such inclination. Some crops were derived from weeds and some weeds were derived from crops, but by far the more usual pattern is the crop–weed complex in which both crop and weed are derived from the same progenitors. Some crops arose in the Vavilovian centers, and others did not; many have centers of diversity, but others do not. Some people were sedentary long before agriculture; others maintained a nomadic way of life long after plants were domesticated and agriculture was established. There is no model with universal, or even very wide application; yet most of them contribute, in some degree, to an understanding of the problem.

One needs to recognize the fact that human beings are enormously varied and their motivations are always complex and never simple. It is difficult enough to psychoanalyze a living, speaking human, so how can we expect to analyze people who lived 10,000 years ago and who belonged to cultures we can but dimly imagine? People do similar things for entirely different reasons and they find very different solutions to the same problems.

A "no-model model" leaves room for whole arrays of motives, actions, practices, and evolutionary processes. What applies in Southeast Asia may not apply at all in Southwest Asia. The patterns in Africa may not be the same as the patterns in Mexico. A search for a single overriding cause for human behavior is likely to be frustrating and fruitless. A humanistic no-model model simply recognizes the likelihood that no single model will explain agricultural origins.

Man did take the initiative in modifying his environment, and plants responded genetically to his activities. He deliberately changed the vegetation with set fires; he sowed seeds; he churned up square miles of land to get tubers, all without developing "agriculture." The development of true agriculture would require more work, but few changes in techniques. It is not even necessary to assume a crisis was always responsible, for the motivations could have been many and various.

The most conspicuous difference between hunting-gathering economies and agricultural ones is in the size of the human populations that can be supported. Farming takes more work, but it can feed more people. Population pressures may or may not have initiated plant domestication, but they have certainly forced the evolution of agricultural economies in a single direction.

Generalizations about human behavior are always hazardous, but there does seem to be a significant difference between agricultural societies and the surviving hunter-gatherers in the role and importance of children. In the agricultural economies, children are an economic asset. They add to the labor force; they create wealth through dowries and bride-prices; and they provide security for the aged. In some societies today, the situation is so intense that childless couples become literally impoverished and may actually starve to death. Even survival sometimes depends upon having children, and the more the better.

The system tends to be self-defeating in the sense that there are strong forces always pressing toward larger populations. More people require more food. More food requires more intensive farming practices which in turn require more work per unit of food. The only way to get more work done is to increase the labor force by having more children. A high value is placed on prolific women and barren ones may be cast out of the society. Subsistence agriculture is not likely to reach equilibrium without external population controls such as disease epidemics, famine, and war.

How far we can push the disequilibrium back toward the beginning of agriculture has not yet been determined. The economic value of children may have been an important influence very early in the evolution of agricultural societies. Certainly, the steady and intense pressures for ever larger populations set into motion trends that are essentially irreversible. Living within the productive capacity of the environment becomes a continual and exhausting struggle. A "hungry time" becomes a part of every year while crop failure means starvation and death. The threat of famine has become a characteristic of agricultural systems; we have no evidence that this was a part of preagricultural systems.

On the other hand, the sample of surviving gatherers is so small and biased that our information may be misleading. The survivors maintain their populations at a fraction of the size that could be supported, but was this true of gatherers in the hearths of agriculture? Perhaps cultivation did begin because of population pressures and degradation of natural resources. How are we to know? Perhaps plant cultivation began in different areas for different reasons.

We have no more facts to support a no-model concept than any other theory, but it does have the advantage of being independent of any set of presuppositions. It is obvious that views of agricultural origins in the past have too often been based on assumptions that have either turned out to be altogether false or that have applied to one situation and not another. The no-model view takes into account the distinct possibility that plant domestication began in different regions for different reasons, and permits us to build theories on evidence as it accumulates rather than on preconceived notions.

The greatest difficulties in understanding agricultural origins trace to a want of information, and no amount of speculation can substitute for evidence. Although we have made some advances in the decades since Darwin wrote that the problem was too difficult to be solved, we are still far from determining the motivation that brought about such a profound change in human adaptation.

Geography of Plant Domestication

No consideration of agricultural origins would be complete without mention of Alphonse de Candolle and N. I. Vavilov. Although neither of them maintained elaborate theories about why or how agriculture originated, they were both concerned about the geography of plant domestication and crop origins.

de Candolle lived in Geneva and was one of the foremost botanists of the 19th century. His book, *Origin of Cultivated Plants* (reprinted in 1959), was primarily an academic and intellectual exercise. He was interested in geography of plants in general and wrote extensively on the subject. He attempted to locate the region of origin of a good many cultivated plants by any means he could. He investigated the distribution of wild relatives, history, names, linguistic derivatives, archaeology, variation patterns, and every other clue he could think of.

In many respects there was not a great deal known in de Candolle's time. Archaeological plant remains were largely confined to materials from the Egyptian tombs and the Swiss lake dwellers. Wild races of a number of plants were not then known, and some of his information was faulty. Nevertheless, his book remains today a model of scholarship and continues to be a useful source of information about the origins of cultivated plants.

Nikolai I. Vavilov was a Russian geneticist and agronomist in charge of an enormous National Institute of Plant Industry. At his disposal were dozens of experiment stations scattered over the Soviet Union, staffed with thousands of professional and sub-professional workers. He proposed one of the most dazzling and ambitious plant-breeding programs ever attempted. It was his plan to collect and assemble all of the useful germplasm of all crops that had potential in the Soviet Union, to study and classify the material, and to utilize it in a national plant-breeding effort. A vigorous, worldwide plant exploration program was launched, and for the first time a systematic plan for genetic resource management was established.

Vavilov was interested in origins because he was interested in genetic diversity, and he thought the two were related. In 1926, he wrote an essay, dedicated to Alphonse de Candolle, *On the Origin of Cultivated Plants* (Vavilov, 1926) in which he proposed that one could reliably determine the center of origin of a crop by an analysis of patterns of variation. The geographic region in which one found the greatest genetic diversity was the region of origin. This was especially true if much of the variation was controlled by dominant genes and if the region also contained wild races of the crop in question.

In this essay, he proposed eight centers of origin with some subcenters, Figure 2.1, and these are widely accepted even today. Actually, much of the plant exploration conducted by his institute had yet to be done and analyses of previous expeditions had not been completed. The work was more of a literature review and expression of philosophical doctrine than a scientific paper based on research data. The techniques for measuring diversity in those days were based on old-fashioned elementary taxonomy. Later, he did develop a classification of agro-ecological groups using such traits as response to day length, cold requirements, reaction to disease, and general adaptation to specific environments.

While data to support his center of origin theory were not provided at the time, an enormous amount of information was generated by the Institute (now called the N. I. Vavilov Institute of Plant Industry or VIR), and published in the *Bulletin of Applied Botany and Plant Breeding* from about 1920 to 1940. The studies at the Vavilov Institute are old now, but when a student at the Crop Evolution Laboratory, University of Illinois, wished to study a crop, we always advised that he or she turn first to the Vavilov Institute publications. "First, see what the Russians said about the crop and go on from there; that is the place to begin." The different crops were studied by professionals who knew their material well and had field experience with it. Many of these studies could not now be conducted because of recent changes in cultivar and landrace usage.

In the past, analyses of world collections or parts of world collections were made and published only in major crops for which large collections, molecular tools, and sufficient funding were available. However, with the advent of affordable next-generation sequencing technologies, most crop species have had at least initial studies of diversity, relationships, and origins performed with high numbers of SNP (single nucleotide polymorphic) markers. Modern studies of diversity have begun to sort out centers of origin and of diversity, and many confirm or revise the intuitive geographic patterns described by Vavilov. These studies often disagree with centers of origin, as diversity patterns depend strongly on type and number of markers or traits used and on sample sizes. Older studies of too few markers or overly specific morphological traits had little or nothing to do with origins. For example, Peeters (1988) used the Cambridge barley collection, recording 12 qualitative and 18 quantitative traits averaged over 3 years, for more than

Figure 2.1 The eight-centers of origin, according to N. I. Vavilov. *Source:* Harlan (1971). Reprinted with permission from AAAS.

100,000 observations and concluded the greatest diversity in barley is in the United States, and very little in Afghanistan and Ethiopia. There was no real center with geographic integrity. Newer and more comprehensive studies with neutral genetic markers can clearly show the center of origin for a crop such as sunflower (Wills & Burke, 2006); however, such comprehensive studies sometimes also show that high genetic diversity within a specific geographical location is not always correlated with the center of domestication of a crop (Burger et al., 2008).

This is due to several reasons. High levels of diversity can be the outcome of secondary contact between crops and their wild relatives which are not their direct progenitors, but still close enough for successful hybridization; this generally happens outside the center of origin (Elstrand et al. 1999; Weissmann et al., 2005). In addition, movement of a crop to new environments via migration or trade, which was common with new and successful domesticates, increases the diversity of the species as it is selected for adaptation to its new environment (e.g., Bedoya et al., 2017; Crites, 1993). Finally, multiple domestication centers have since been identified for some crops (Brown et al., 2009; Civan et al., 2015).

Vavilov predicted some of these events when he had to concede that his method of "differential phytogeography" did not work very well. He invented the concept of secondary centers to account for the fact that centers of diversity are not the same as centers of origin. In fact, the variation in secondary centers is often much greater than in the centers of actual domestication. He also developed the concept of secondary crops to those that were derived from weeds of older, primary crops. Rye and oats were cited as examples. As agriculture spread from the Near East and Mediterranean centers toward northern Europe, weed rye and weed oats were carried along as contaminants of the barley and emmer fields. In due course domesticated races developed, far removed from the original homeland of rye and oats. As we have seen, Edgar Anderson (1954) favored the idea that crops were often derived from weeds and he was strongly influenced by Vavilov's writings.

The concept of center of origin has evolved since Vavilov's time. Basically, what Vavilov did was to draw lines around areas in which agriculture has been practiced for a very long time and in which indigenous civilizations arose. The geography of crop variation depends a lot upon the geography of human history.

When one actually analyzes origins crop by crop, it soon becomes apparent that many of them did not originate in Vavilovian centers. Some crops do not even have centers of diversity. The pattern is much more complex and diffuse than Vavilov had visualized. In the case of the Near East, we seem to have a definable center in the sense that a number of plants and animals were domesticated within a relatively small region and were diffused outward from the center. In Africa, nothing of the sort is apparent. The evidence seems to indicate that activities of plant domestication went on almost everywhere south of the Sahara and north of the equator from the Atlantic to the Indian Ocean. Such a vast region could hardly be called a "center" without distorting the meaning of the word, so I called it a noncenter (Harlan, 1971). In North China, there seemed to be fairly convincing evidence for a center, but nothing of the sort is evident in Southeast Asia and the South Pacific. The pattern may be similar in the Americas with a center in Mesoamerica and a noncenter in South America. My own version of agricultural origins is shown in Figure 2.2.

I have proposed three independent systems, each with a center and a noncenter. I also visualized some stimulation and feedback in terms of ideas, techniques, or materials between center and noncenter within each system. Since making these proposals, my centers have been eroding by more information. The Near Eastern "center" is flanked by activities in the Caucasus (Lisitsina, 1984) and possibly the Balkans and Ethiopia. For animal domestication, the Near Eastern "center" is flanked by domestication in Baluchistan, Europe, and Africa. The Chinese "center" has become much more diffuse than it once seemed. After the beginning of the Holocene, a mosaic of Mesolithic cultures evolved over most of China, and from these several Neolithic cultures developed (Chang, 1986). The pattern now appears to be a mosaic of developments over a broad front rather than one of a small, restricted center in which innovations occurred and out of which they were diffused. The Mesoamerican "center" is mosaicked by independent developments in the mid-Mississippi-lower Ohio watersheds (Smith, 1989), in Sonora, Arizona (Ford, 1981, 1985), and northeast Mexico. With respect to origins of agriculture, it is perhaps time to abandon the concept of centers of origin altogether. Individual crops may or may not have centers of origin and many have centers of diversity, but agriculture as a food-procurement system has no specific time or place of origin. In the geographic sections of the book to follow, I shall refer to regions rather than centers.

An Ecological Approach

The geography of domestication might make more sense if we examined the ecological settings to see what conditions are most likely or most unlikely to be suitable for agricultural origins. This, of course, is what

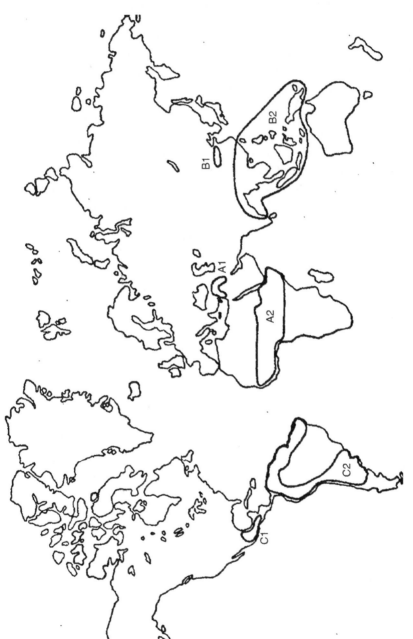

Figure 2.2 Centers and noncenters of agricultural origins: (A1), Near East center, (A2), African noncenter, (B1), North Chinese center, (B2), Southeast Asian and South Pacific noncenter, (C1) Mesoamerican center, and (C2) South American noncenter. *Source:* Harlan (1971). Reprinted with permission from AAAS.

C. O. Sauer tried to do, but in this treatment, I shall make use of experience and hindsight. We may list the major climate or vegetation formations as:

1) Tundra and taiga
2) Temperate forests
3) Temperate prairies
4) Temperate steppes
5) Mediterranean woodlands
6) Tropical forest
7) Tropical savanna
8) Deserts
9) Tropical highlands
10) Sea coasts

The tundra and taiga can easily be ruled out. To this day we have not been able to do much with them agriculturally. Reindeer were domesticated and some forestry is practiced, but there is little in the way of farming. The well-developed temperate prairies (Figure 2.3) can be ruled out as being too difficult for primitive tools. They were not developed in North America nor in the Ukraine and Russia until special steel plows were invented to turn the sod. Some of our most productive soils could not be exploited until drainage was developed as well. Native peoples of the North American prairie region who practiced agriculture kept to wooded loess soils of river terraces and woodland bottoms and avoided the prairie proper. White settlers who followed did the same, moving from woodland to woodland and skipping the prairie because they could not manage the sod. The only major crop that might be ascribed to temperate prairie is sunflower (*Helianthus annuus* L.) and it was initially cultivated in adjacent woodlands.

The steppes tend to be marginal, and they are an unlikely place to begin. The few domestic plants that might be of steppe origin are *Panicum miliaceum*, *Setaria italica* (L.) P. Beauv., and *Cannabis sativa* L. One invaluable contribution, however, came from *Aegilops squarrosa* L., a plant of the central Asian steppes that provided the D genome of hexaploid wheat. This may be why wheat can be grown on such a scale on the temperate steppes of the world.

The tropical rain forests (Figure 2.4) provide a very difficult environment. We do not know of a single hunting-gathering society of rain forests that does not require some supplementation from cultivated plants (Bailey et al., 1989). We have no evidence of early occupation of either the Amazonian or African rain forest. Certainly, the present rain forest of Amazonia has been strongly modified by activities of farming people and does not represent the original conditions faced by the first colonizers. It is probable that agriculture must first be developed and adapted to this

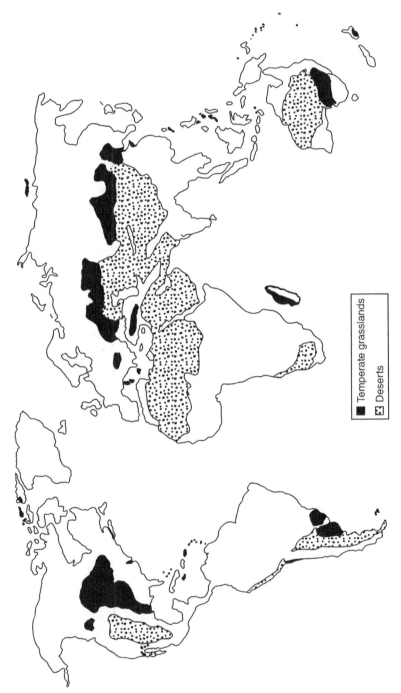

Figure 2.3 Well-developed temperate prairies and deserts.

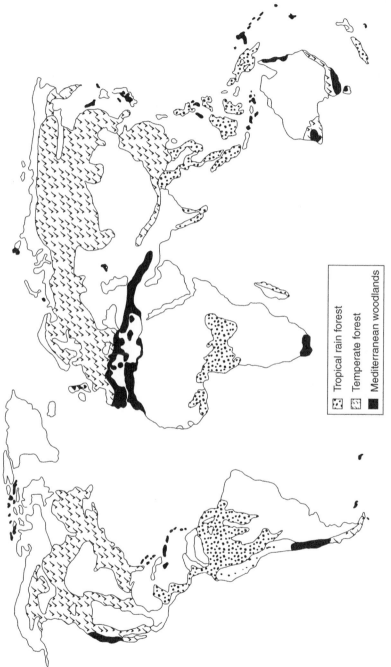

Figure 2.4 Tropical rainforests, temperate forests, and Mediterranean woodlands.

Table 2.2   The world's 30 leading food crops in terms of estimated edible dry matter.[†]

Crop	MT[‡]	Annual (A) or perennial (P)	Ecological origins[§]	Self (S) or cross (C) fertilizing[ᵛ]	Ploidy	Chromosome no(s).
Cane Sugar	1841.5	P	R	–/C	many	40 – 128
Maize	1134.7	A	S	C	2x	20
Wheat	771.7	A	M	S	2x, 4x, 6x	14, 28, 42
Rice	769.7	A	S	S	2x	24
Potato	388.2	A	H	–/C	2x, 4x, 6x	24, 48, 72
Soybean	352.6	A	W	S	4x	40
Cassava	292.0	P	S	–	2x	36
Beet Sugar	301.0	A	C	C	2x, 3x, 4x	18, 27, 36
Tomato	182.3	A	C	S	2x	24
Barley	147.4	A	M	S	2x	14
Banana	113.9	P	R	–	3x	33
Sweet Potato	112.3	A	S	–C	6x	90
Onion	97.9	A	?	C	2x	16
Apple	83.1	P	W	–C	2x, 3x	34, 51
Rapeseed	76.2	A	M	C	4x	38
Grape	74.3	P	W	–	2x	38
Orange	73.3	P	R	–/C	2x, 3x	18, 27
Yams	73.0	A	S	–	2x – 14x	20 – 140
Cabbage	71.5	A	C	C	2x	18
Coconut	60.8	P	C	C	2x	32
Sorghum	57.6	A	S	S	2x	20
Mango	50.6	P	R	–/C	2x	40
Sunflower	47.7	A	P	C	2x	34
Peanut	47.1	A	S	S	4	40
Pea (dry & Green)	36.9	A	M	S	2x	14
Millets	28.5	A	S	S/C	2, 4	18, 36
Bean (dry & green)	27.3	A	S	S	2x	22
Oats	25.9	A	M	S	6x	42
Peaches & Nectarines	24.7	P	–	–	2x	16
Rye	13.7	A	M	C	2x	14

Table 2.2   (Continued)

Plant Food Crop Group	Total (addition of above)
Cereals	2949.2 MT
Tubers	1093.4 MT
Pulses	463.9 MT
Fruits	77.1 MT
Sugar	2142.5 MT
Annuals	4763.2 MT
Perennials	2615.2 MT
Mediterranean and savanna (long dry seasons)	884.9 MT

Animal Food Group	
Edible dry matter of meats (57% pig; 31% cattle)	209.0 MT
Edible dry matter of poultry (93% chicken)	116.0 MT
Edible dry matter of milk	794.4 MT
Edible dry matter of eggs	80.1 MT

[†] Production from FAO Production Yearbook, 2017; www.fao.org/faostat/en/#data/OC (accessed 2/6/2019).
[‡] MT=million metric tons.
[§] M=Mediterranean, S=savanna, W=woodlands, R=tropical forest, H=highlands, C=coastal, and P=prairie.
[¥] – = vegetative propagation

difficult environment before people could live in it yearlong. Of the major crops listed in Table 2.2, those that might have come from a rain forest environment are: sugarcane, bananas and plantains, and orange and mango, but these are not plants of the closed canopy. They are adapted to the forest margins, stream banks, and modified forests where they can receive more sunlight than in a high closed canopy forest. This tends to be true of other products of tropical forest. The formation as a whole has yielded a large number of useful plants, mostly fruits and nuts and some of these will be mentioned in later chapters. While the forest-savanna eco-tone is rich in potential, the forest itself is an inhospitable place to begin agriculture.

Deserts (Figure 2.3) have some possibilities, if there is water available. The Sonoran complex evolved with local domestication of the tepary bean, devil's claw, and *Panicum hirticaule*. Other parts of the complex, maize, squash, beans, and cotton were presumably obtained from farther south in Mexico. In Africa and the Near East, the date was a major contribution from the desert environment and pearl millet was probably domesticated in the Sahara. *Prosopis, Acacia, Zizyphus, Borassus*, and other trees have been heavily exploited if not domesticated. No desert crop has made the select list and the environment is generally a very unlikely one for the beginnings of agriculture.

Temperate forests (Figure 2.4) have somewhat better possibilities. Clearings can be made by deadening trees. The soils of loess terraces, at least, are friable and easily worked with primitive tools, and leaf mold and litter can be helpful in soil conditioning. The contributions, as one would expect, have been primarily in fruits and nuts, for example, apple, pear, peach, cherry, quince, plum, grape, walnut, hickory, pecan, hazelnut, chestnut, buckeye, oak, and so forth; the last two usually requiring detoxification. A small complex developed in eastern North America where *Iva, Chenopodium, Phalaris, Polygonum, Ambrosia, Hordeum pusillum* Nutt., possibly a *Cucurbita* and sunflower were domesticated (Watson, 1989). Still, cultivation of such plants seems to be late and they continued to be gathered from the wild long after other crops had been domesticated. On the whole, temperate forests are benign environments and agriculture was unnecessary until rather late in prehistory; see comments on Jo-mon of Japan, Chapter 10.

A look at Figure 2.4 suggests why the Near East appears to have a center of origin. The region of winter rainfall and summer drought is relatively small, and the area where distributions of wild wheat, barley, sheep, and goats overlap is even smaller. This climatic regime and vegetative formation occurs on the western side of land masses between 30° and 40° north and south. The largest area of this climatic regime and associated vegetation is, therefore, around the Mediterranean and fanning eastward into the deserts of Iran and Afghanistan. The portion of the area with adequate rainfall for good development of wild cereals and consequently of dryland farming is restricted to an arc along the Zagros and Tauros mountains at mid-elevations and down the Levant to a little south of Jerusalem. These last zones are restricted to the east by high mountains. We probably have a "center" because it could not have happened any other way. This did not exclude the possibilities of independent developments elsewhere. Similar climates exist in other regions such as South Africa and southern Australia which just reach the zone, while California and Chile intercept the full width.

On the list of 30 major food crops, the Mediterranean region contributed: wheat, barley, pea, rapeseed, and the wild races of oats and rye. The annual production of these five crops is on the order of 1072 million metric tons.

The savanna formation includes both open grasslands with widely scattered trees and dry forest where the dry season lasts some 5–8 months each year. The regime favors both seedy annuals and plants with tubers that behave like annuals. Tuber formation is an adaptation to long dry seasons. At the onset of the rains, the tubers of yams, for example, sprout and the vines grow with remarkable vigor; virtually the entire contents of the tuber are mobilized and translocated upward. At the end of the rains, the process is reversed and nearly all of the food stored in the vine is translocated downward and the tuber grows very rapidly. The vine dies and the tuber remains dormant through the dry season and safe from fires that often burn the vegetation at that time.

The annual habit is also suited to long dry seasons. Seeds can survive the drought and sprout at the onset of the rains. The annual races of wild rice grow in waterholes that are filled with water during the rains and dry up in the dry season. Wild maize (teosinte) is adapted to open dry forest at mid-elevation in Mesoamerica. On our select list of food crops, the savanna and dry forests can claim: maize, rice, sorghum, cassava, sweet potato, bean, peanut, yams, cotton (seed oil) with a cumulative annual production of some 2519 million metric tons.

Tropical highlands (Figure 2.5) have yielded some major crops and many minor ones. An important suite of crops evolved in the Andes. On the world scene, the important one is potato, but others are very important locally. Some are listed in Table 3.1 (Chapter 3), and mentioned in Chapter 11. Economically, the most important contribution of the East African highlands is *arabica* coffee. This is not a food crop, but generates a lot of money that can buy food, so it is important on the world scene. Other crops of the region are treated in Chapter 9.

The sea coasts of the world have provided some important crops. On the select list, these include coconut, cabbage, and beet. Radish and a few others can be added. The coconut may have some considerable antiquity as a cultivated plant, but the others appear to be rather late.

Seen from an ecological perspective, early agriculture could have evolved in a variety of settings, but the greatest opportunities would be in regions with long dry seasons where a wide selection of annual seed crop progenitors and seasonal tuber crop progenitors was available.

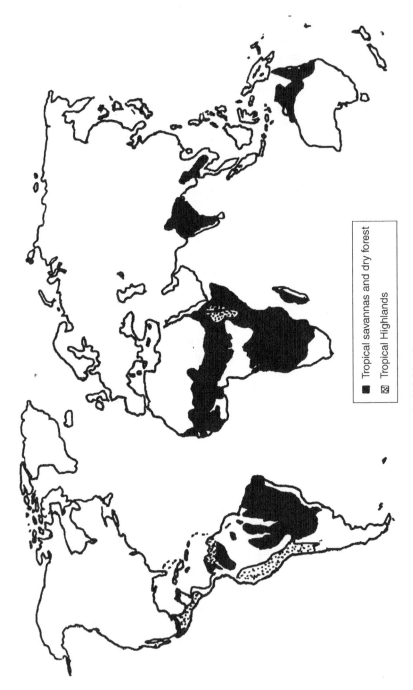

Figure 2.5 Tropical savannas and dry forests, and tropical highlands.

Conclusions

It is very unlikely that one model would explain all situations for plant domestication. There are too many independent beginnings for that. One scenario that is likely is one in which people of a well-developed delayed return hunter-gathering society began to grow one or a few special species in gardens, perhaps for fun, perhaps for convenience, perhaps to bridge a lean time in the gathering schedule, but more likely, to raise a chosen plant in a spiritually safe space free of malevolent forces.

Such a scenario would be as much a nonevent as the Kuruk growing tobacco. The change would be completely trivial until and unless the early initiative of small-scale gardening evolved into true food production, and this might take millennia. It may be that agriculture slipped through the back door without anyone noticing. This scenario seems to fit the evidence from Oaxaca, the Andes, and the midwestern USA. Probably other scenarios were played out in the Old World.

If this view of American "neolithization" is more or less correct, then the major changes and adjustments of human adaptation came *before* plant cultivation and we should be looking at what motivated changes in the Epipaleolithic or Mesolithic. What prompted people all over the world to make smaller, more elegant and more efficient tools and weapons? What prompted them to take to the water in rafts, canoes, boats, make harpoons, fish hooks, nets, traps, weirs, and so forth? What motivated a broader spectrum of hunting and gathering? Here, we do not have to look far for incentives. With all that ice melting and sea levels rising, it was a watery world, and with mass faunal extinction, other resources had to be exploited. After the adjustments were made, the best opportunities for initiation of plant cultivation would be in areas with long dry seasons, whether temperate or tropical.

References

Anderson, E. (1954). *Plants, man, and life*. London: A. Melrose.

Angel, J. L. (1984). Health as a crucial factor in the changes from hunting to developed farming in the eastern Mediterranean. In M. Cohen & G. J. Armelagos (Eds.), *Paleopathology at the origins of agriculture* (pp. 51–78). New York, NY: Academic Press.

Bailey, R. C., Head, G., Jenike, M., Owen, B., Rechtman, R. R., & Zechenter, E. (1989). Hunting and gathering in tropical rain forest: Is it possible?

American Anthropologist, 91, 59–82. https://doi.org/10.1525/aa. 1989.91.1.02a00040

Bedoya, C. A., Dreisigacker, S., Hearne, S., Franco, J., Mir, C., Prasanna, B. M., . . . Warburton, M. L. (2017). Genetic diversity and population structure of native maize populations in Latin America and the Caribbean. *PLoS One, 12*(4), e0173488. https://doi.org/10.1371/journal. pone.0173488

Berndt, R. M., & Berndt, C. (1970). *Man, land and myth in north Australia: The gunwinnggu people*. East Lansing, MI: Michigan State University Press.

Binford, L. R. (1968). Post-Pleistocene adaptations. In S. R. Binford & L. R. Binford (Eds.), *New perspectives in archaeology* (pp. 313–341). Chicago, IL: Aldine.

Black, J. N. (1971). Energy relations in crop production—a preliminary survey. *The Annals of Applied Biology, 67*, 272–278. https://doi. org/10.1111/j.1744-7348.1971.tb02928.x

Boserup, E. (1965). *The conditions of agricultural growth*. London: Allen & Unwin.

Braidwood, R. J. (1972). Prehistoric investigations in southwestern Asia. *Proceedings of the American Philosophical Society, 116*, 310–320.

Brown, T. A., Jones, M. K., Powell, W., & Allaby, R. G. (2009). The complex origins of domesticated crops in the Fertile Crescent. *Trends in Ecology & Evolution, 24*, 103–109. https://doi.org/10.1016/j.tree.2008.09.008

Burger, J. C., Chapman, M. A., & Burke, J. M. (2008). Molecular insights into the evolution of crop plants. *American Journal of Botany, 95*, 113–122. https://doi.org/10.3732/ajb.95.2.113

Carter, G. F. (1971). Pre-Columbian chickens in America. In C. L. Riley (Ed.), *Man across the sea* (pp. 178–218). Austin, TX: University of Texas Press.

Chang, K.-C. (1986). *The archaeology of ancient China* (4th ed.). New Haven, CT: Yale University Press.

Chase, A. K. (1989). Domestication and domiculture in northern Australia: A social perspective. In D. R. Harris & G. C. Hillman (Eds.), *Foraging and farming: The evolution of plant exploitation* (pp. 42–54). London: Unwin Hyman.

Chavannes, E. (1967). *Les mémoires historiques de Se-Ma Ts'ien*. Paris, France: Adrien-Maisonneuve.

Childe, V. G. (1925). *The dawn of European civilization*. New York, NY: Alfred Knopf.

Childe, V. G. (1952). *New light on the most ancient east*. London: Routledge & Paul.

Christie, A. (1983). *Chinese mythology*. Fetham, England: Hamlyn Publishing Group.

Civáň, P., Craig, H., Cox, C. J., & Brown, T. A. (2015). Three geographically separate domestications of Asian rice. *Nature Plants, 1*, 15164. https://doi.org/10.1038/nplants.2015.164

Cohen, M. (1977). *The food crisis in prehistory*. New Haven, CT: Yale University Press.

Cohen, M. N., & Armelagos, G. J. (1984). *Paleopathology at the origins of agriculture*. New York, NY: Academic Press.

Crites, M. A. (1993). Domesticated sunflower in 5th millennium BP temporal context—new evidence from Middle Tennessee. *American Antiquity, 58*, 146–148. https://doi.org/10.2307/281459

Cummins, J. S. (1962). *The travels and controversies of Friar Domingo Navarrette 1618–1686* (Vol. *1*). Cambridge, UK: Hakluyt Society at the University Press.

Darwin, C. (1896). *The variation of animals and plants under domestication* (2nd ed.). New York, NY: D. Appleton & Co.

Darwin, C. (1909). *The descent of man and selection in relation to sex* (2nd ed.). New York, NY: D. Appleton & Co. https://doi.org/10.5962/bhl.title.70714

de Candolle, A. (1959). *Origin of cultivated plants* (2nd ed.). New York, NY: Hafner. Translated from the 1886 edition.

de la Vega, G. (1961). *The royal commentaries of the Inca Garcilaso de la Vega*. New York, NY: Orion.

Dorson, R. M. (1972). *African folklore*. Bloomington, IN: Indiana University Press.

Eberhard, W. (1965). *Folktales of China*. London: Routledge & Kegan Paul.

Ellstrand, N. C., Prentice, H. C., & Hancock, J. F. (1999). Gene flow and introgression from domesticated plants into their wild relatives. *Annual Review of Ecology and Systematics, 30*, 539–563. https://doi.org/10.1146/annurev.ecolsys.30.1.539

Evans, L. T. (1975). The physiological basis of crop yield. In L. T. Evans (Ed.), *Crop physiology: Some case histories* (pp. 327–355). Cambridge, UK: Cambridge University Press.

FAO (2017). *FAO production yearbook*. Rome, Italy. Retrieved from http://www.fao.org/faostat/en/#data/OC: FAO. (accessed 6 Feb. 2019)

Fiore, S. (1965). *Voices from the clay*. Norman, OK: University of Oklahoma Press.

Fitzgerald, C. P. (1950). *China, a short cultural history*. New York, NY: D. Appleton-Century Co.

Flannery, K. V. (1968). Archaeological systems theory and early Mesoamerica. In B. J. Meggers (Ed.), *Anthropological archaeology in the Americas* (pp. 67–87). Washington, DC: Anthropological Society of Washington.

Flannery, K. V. (1986). *Guilá Naguitz: Archaic foraging and early agriculture in Oaxaca, Mexico*. New York, NY: Academic Press.

Ford, R. I. (1981). Gardening and farming before A.D. 1000: Patterns of prehistoric cultivation north of Mexico. *Journal of Ethnobiology, 1,* 6–27.

Ford, R. I. (1985). *Prehistoric food production in North America. Anthropological Paper no. 75.* Ann Arbor, MI: Museum of Anthropology, University of Michigan.

Fox, W. S. (1916). Greek and Roman. In L. H. Gray (Ed.), *Mythology of all races.* Boston, MA: Marshall Jones Co.

Hahn, E. (1896). *Die Haustiere und ihre Beziehungen zur Wirtschaft des Menschen.* Leipzig, Germany: Duncker und Humbolt.

Hahn, E. (1909). *Die Entstehung der Pflugkultur.* Heidelberg, Germany: C. Winter.

Harlan, J. R. (1956). Distribution and utilization of natural variability in cultivated plants. In: Brookhaven National Laboratory (ed.). In *Genetics in plant breeding, Brookhaven Symposia in Biology* (pp. 191–206). New York, NY: Brookhaven National Laboratory.

Harlan, J. R. (1961). Geographic origin of plants useful to agriculture. In R. E. Hodgson (Ed.), *Germplasm resources, American Association of Advancement of Science and Publicaton no. 66* (pp. 3–19). Washington, DC: American Association of Advancement of Science.

Harlan, J. R. (1971). Agricultural origins: Centers and noncenters. *Science, 174,* 468–474. https://doi.org/10.1126/science.174.4008.468

Harlan, J. R. (1980). Origins of agriculture and crop evolution. In M. K. Harris (Ed.), *Biology and breeding for resistance to arthropods and pathogens in agricultural plants* (pp. 1–8). College Station, TX: Texas Agricultural Experiment Station.

Harlan, J. R. (1986). Plant domestication: Diffuse origins and diffusion. In C. Barigozzi (Ed.), *The origin and domestication of cultivated plants* (pp. 21–34). Amsterdam, The Netherlands: Elsevier. https://doi.org/10.1016/B978-0-444-42703-8.50007-5

Harlan, J. R., & Zohary, D. (1966). Distribution of wild wheats and barley. *Science, 153,* 1074–1080. https://doi.org/10.1126/science.153.3740.1074

Healey, C. J. (1988). Culture as transformed disorder: Cosmological evocations among the Maring. *Oceania, 59,* 106–122. https://doi.org/10.1002/j.1834-4461.1988.tb02314.x

Langlie, B., Mueller, N. G., Spengler, R. N., & Fritz, G. J. (2014). Agricultural origins from the ground up: Archaeological approaches to plant domestication. *Botany, 101*(10), 1601–1617.

Latourette, K. S. (1941). *The Chinese, their history and culture.* New York, NY: Macmillan Co.

Lee, R. B., & DeVore, I. (1968). Problems in the study of hunters and gatherers. In R. B. Lee & I. DeVore (Eds.), *Man the hunter* (pp. 3–12). Chicago, IL: Adine.

Lisitsina, G. N. (1984). The Caucasus—a center of ancient farming in Eurasia. In W. Van Zeist & W. A. Casparie (Eds.), *Plants and ancient man: Studies in palaeoethnobotany* (pp. 285–292). Rotterdam, The Netherlands: A.A. Balkema.

Mabuchi, T. (1964). Tales concerning the origin of grains in the insular areas of Eastern and Southeastern Asia. *Asian Folklore Studies, 23*, 1–92. https://doi. org/10.2307/1177638

Maurice, T. (1795). *The history of Hindoostan.* New Delhi, India: Novrang.

Mayer, F. H. (1986). *The Yanagita Kunio guide to the Japanese folk tale.* Bloomington, ID: Indiana University Press.

Oldfather, C.H. (Transl.) (1946). Diodorus of Sicily. (Vol 10) Cambridge, MA: Harvard University Press.

Peeters, J. P. (1988). The emergence of new centres of diversity: Evidence from barley. *Theoretical and Applied Genetics, 76*, 17–24. https://doi.org/10.1007/BF00288826

Pickersgill, B. (2007). Domestication of plants in the Americas: Insights from Mendelian and molecular genetics. *Annals of Botany, 100*(5), 925–940. https://doi.org/10.1093/aob/mcm193

Pimentel, D. (1974). *Energy use in world food production. Environmental Biological Report no. 74-1.* Ithaca, NY: Cornell University.

Pimentel, D., & Hall, C. W. (1989). *Food and natural resources.* New York, NY: Academic Press. https://doi.org/10.1016/B978-0-12-556555-4.50006-5

Prescott, W. H. (1936). *History of the conquest of Mexico and history of the conquest of Peru.* New York. NY: Modern Library.

Recinos, A. (1947). *Popol Vuh, Las antiquas historias del Quiché.* Mexico, D.F.: Fondo de Cultura Economica.

Reichel-Dolmatoff, G. (1971). *Amazonian cosmos: The sexual and religious symbolism of the Tukano Indians.* Chicago, IL: University of Chicago Press.

Sauer, C. O. (1952). *Agricultural origins and dispersals.* Cambridge, MA: M.I.T. Press.

Smith, B. D. (1989). Origins of agriculture in eastern North America. *Science, 246*, 1566–1571. https://doi.org/10.1126/science.246.4937.1566

Vavilov, N. I. (1926). *Studies on the origin of cultivated plants.* Leningrad, Russia: Institut de Botanique Appliquée et d'Amelioration des Plantes Cultivees.

Watson, P. J. (1989). Early plant cultivation in the eastern woodlands of North America. In D. R. Harris & G. C. Hillman (Eds.), *Foraging and farming: The evolution of plant exploitation* (pp. 555–571). London: Unwin Hyman.

Weissmann, S., Feldman, M., & Gressel, J. (2005). Sequence evidence for sporadic intergeneric DNA introgression from wheat into a wild *Aegilops* species. *Molecular Biology and Evolution, 22,* 2055–2062.

Wilbert, J. (1987). *Tobacco and shamanism in South America.* New Haven, CT: Yale University Press.

Wills, D. M., & Burke, J. M. (2006). Chloroplast DNA variation confirms a single origin of domesticated sunflower (*Helianthus annuus* L.). *The Journal of Heredity, 97,* 403–408. https://doi.org/10.1093/jhered/esl001

3

What Is a Crop?

The fountain which from Helicon proceeds, that sacred stream, should never water weeds, nor make the crop of thorns and thistles grow.
Roscommon by Wentworth Dillon (Johnson, 1827)

It is not always easy to distinguish between wild and cultivated plants in South America, and there are many intermediate stages between the utilization of plants in their wild state and their true cultivation.
Lévi-Strauss (1950)

Source: Chapter title page sketches by Patricia J. Scullion, ASA/CSSA Headquarters Office.

Harlan's Crops and Man: People, Plants and Their Domestication, Third Edition.
H. Thomas Stalker, Marilyn L. Warburton, and Jack R. Harlan.
© 2021 American Society of Agronomy, Inc. and Crop Science Society of America, Inc.
Published 2021 by John Wiley & Sons, Inc.
doi:10.2135/harlancrops

Definitions

According to unabridged dictionaries, the word "crop" has several meanings. One set of definitions involves the verbal form of the concepts of cutting, mowing, grazing, lopping off branches, and so on. Sheep crop grass closely; a head of hair or a mane of a horse is cropped. Other definitions involve the material that is harvested, whether it be plant or animal. The forester may speak of a timber crop, the livestock man of a calf crop or a lamb crop. The material harvested is referred to as a "crop." In other cases, "crop" specifies certain kinds of plants that are grown on purpose for a later harvest. Even so, it would be quite appropriate for a Native American to speak of a "wild-rice crop." Note that in the poem quoted above, Roscommon speaks of a crop of thorns and thistles. This is probably not what the Crop Science Society of American had in mind when it adopted its name.

It is perhaps appropriate that the term "crop" is broad and somewhat ambiguous because many of the plants we grow for food are not fully domesticated and the word "crop" covers all that which is harvested regardless of its status as a domesticate. We must therefore make the distinction between "cultivated" and "domesticated" as clearly as possible. The terms are often used synonymously, but actually they have quite different implications.

The words "domestication" and "to domesticate" are derived from the Latin domus, house, dwelling, household. To domesticate means to bring into the household. A domestic is a servant who lives in the house. A domesticated plant or animal is one that has been brought into the household and serves those who also live there. The household, then, can include the home garden, barnyard, sty, fold, field, orchard, or ranch. The cattle tribes of Africa live by their domesticated animals, and their domuses move from camp to camp and from kraal to kraal. The overall grazing range may be considered a domain (derived from dominus, lord, not domus). In like manner, among Aborigines, the total foraging range may be called a domain, within which are more discrete hearth-based domuses (Chase, 1989). A domus is more than space or territory; it is an area intimately known and spiritually safe. The imprint of man is on it and the demons and spirits are benign. Here one can be "at home."

Australian anthropologists, in particular, have been groping for words derived from domus to describe the interactions of Aborigines and their habitats. The word "domiculture-household economy" has been revived and words like "domisticatory" coined. This has been prompted by the concept that while the Aborigines did not domesticate any plants or animals, they did domesticate the environment in which they live. The landscape can be thought of as being brought into the household.

Rindos claimed that plants were domesticated before agriculture as a result of a long coevolution with man. In fact, he wrote: ". . . the indehiscent rachis of the small grains is as much the cause as the result of agriculture" (Rindos, 1984, p. 139). The argument has little merit in view of the fact that wild grass seeds with fragile rachises can be harvested in commercial quantities and compete with domesticated cereals in the market place (Harlan, 1989). The concept, however, is interesting from the point of view of the meaning of domestication. The consensus is that domestication involves genetic changes that adapt the plant or animal to the domus, and full domestication results in populations that cannot survive without the aid of man. Cereals with indehiscent rachises and legumes with indehiscent pods are domesticated. The vegetation of "domesticated landscapes" can survive without the aid of man, but the modified landscapes cannot. And yet, some 35,000–40,000 years of landscape domestication in Australia has not resulted in domesticated plants. Plants were clearly not domesticated before agriculture. The term should be used with more precision.

Since domestication is an evolutionary process, there will be found all degrees of plant and animal associations with man and a range of morphological differentiations from forms identical to wild races to fully domesticated races. A fully domesticated plant or animal is completely dependent on man for survival. Therefore, domestication implies a change in ecological adaptation, and this is usually associated with morphological differentiation. There are inevitably many intermediate states.

To cultivate means to conduct those activities involved in caring for a plant, such as tilling the soil, preparing a seedbed, weeding, pruning, protecting, watering, and manuring. Cultivation is concerned with human activities, while domestication deals with the genetic response of the plants or animals being tended or cultivated. It is therefore quite possible to cultivate wild plants, and cultivated plants are not necessarily domesticated.

Harvested plant materials may be classified as wild, tolerated, encouraged, or domesticated. We have shown in Chapter 1 that a very large number of species has been harvested in the wild, not only by gatherers but by fully established cultivators as well. Examples of tolerated and encouraged plants will be given in the following section.

Intermediate States

Lévi-Strauss' (1950) observation on the distinction between wild and cultivated plants (quoted at the beginning of this chapter) need not have been restricted to South America. Although the situation widely occurs in the tropics, it applies to many temperate crops as well. We shall examine below a few examples of intermediate states between wild and domesticated plants.

Baobab Tree

There is a class of plants that came to be closely associated with man, but without evident genetic modifications. One example is the baobab (*Adansonia* spp.), a tree widely distributed through the savannas of Africa, South Asia, and northern Australia. The plant can become very large and is useful in many ways: the fruit can be eaten, the leaves make good pot-herbs, the bark supplies fiber, and the great hollow boles can be used to store water. In Africa, there is a fairly close association between village sites and stands of baobab. To what extent the villages are located near baobabs by design or how frequently the trees become established after the village is founded we do not know. Certainly, if fruits are repeatedly brought to a village, it would seem that new seedlings would inevitably become established, and since they are recognized and protected, stands would develop.

Acacia

Another example is the case of white-barked acacia (*Acacia albida* Del.). The natural habitat of this tree appears to be along ephemeral watercourses in the dry savanna, but vast stands have become established over tens of thousands of square kilometers in areas where it would not naturally occur. It is a dominant species of certain manmade landscapes of Africa. The tree has the peculiar habit of shedding its leaves at the start of the rains and going dormant through the rainy season. For this reason, it does not compete with interplanted crops and the cultivators believe that crops yield more in association with the tree than without it. Some agronomic studies indicate that this belief is correct (Dancette & Poulain, 1968). At any rate, the tree has prospered enormously by the selective protection of man without any apparent genetic modifications toward domesticated races.

Karité Tree

The karité or shea butter tree [*Butyrospermum parkii* (G. Don) Kotschy] is a similar case, except that protection is reinforced by superstition or religious feelings. An edible oil is extracted from the fruit. The plants are individually owned, considered valuable, and are almost never cut down. Other trees of the savanna may be cut for firewood, charcoal, house construction, and other uses, but the karité enjoys the status of a semisacred tree. As a result, vast areas of the broadleaved savanna of West Africa are covered with nearly pure stands of evenly spaced karité trees.

Oil Palm

The West African oil palm (*Elaeis guineensis*) is an even more elaborate model in some ways. Wild stands occur near the edges of the forest, but the plant is not sufficiently tolerant to deep shade for it to grow in dense forest. As shifting cultivation has reduced the high forest to bush, however, the oil palm has invaded the forest zone. In the process of shifting cultivation, the farmers slash the bush during the dry season and burn it, reducing the vegetation sufficiently that one or two crops may be grown in the burned area. The oil palm, however, is spared. As a result, the palm is encouraged, and over a period of years, stands become thicker and thicker. In some areas, very extensive stands of oil palms developed without anyone ever purposely planting a seed.

The palm is saved from cutting because it is a valuable tree for seed oil, but this procedure is reinforced by local tradition. There is a belief in some tribes that if an oil palm is cut down, there will be a death in their village, so no one among these tribes cuts it.

The fruits of the oil palm are borne in large bunches. At maturity, the bunch is cut and fruits the size of plums are beaten off with sticks. There is a belief among some tribes that if this is done in a village, the flying fruits represent people leaving the village. As a consequence, the fruits are beaten off somewhere in the bush, not in the village, thus helping to disseminate the species.

Finally, there is a gene in the oil palm that controls the development of the kernel inside the fruit. One allele in the homozygous condition produces a kernel with a very thick shell called the durra type. The other allele in homozygous conditions produces no kernel and, this type is called pisifera. The heterozygote is called tenera and produces a thin-shelled kernel. The pisifera, having no kernel, is female-sterile, and the gene frequency for the allele would tend to decline except that the people prefer to harvest the tenera and pisifera plants. There is a tendency to tap the durra plants for palm wine instead of harvesting them for oil. Repeated tapping kills the tree, reducing the frequency of the durra genotype and raising the frequency of tenera and pisifera trees.

Here we have a plant that is encouraged, disseminated, harvested, and selected without anyone deliberately planting a seed. Is the oil palm in indigenous agriculture a cultivated plant or not? In the 20th century, it has become a very important plantation crop in the wet tropics; its hectarage is increasing; and the yields of new hybrids are very high. Under plantation conditions, the high-yielding hybrids are domesticated races, but under traditional systems, the status of the plant is very different.

Sago Palm

Another example of an intermediate step between wild and domesticated involves the sago palm (*Metroxylon sagu* Rottb.). In parts of Melanesia, the sago palm is an important item in the diet. To harvest the tree, it is cut down and split open; the interior is full of a starchy pith, which is edible. There are two kinds of sago: one that is very thorny and difficult to handle and another in which the thorns are absent. It is a simple protoagricultural practice to cut out the thorny trees when they are young and thereby develop a pure stand of the smooth types. Again, man is selecting among wild populations.

Ethiopian Oats

Ethiopian oats may be cited as an example of a tolerated species. These rather strange oats are related to the tetraploid *Avena barbata* of the Near East and Mediterranean regions. They arrived in Ethiopia as weeds in barley and emmer fields. Although the Ethiopian cultivators do not grow oats as a separate crop, they do tolerate a mixture of oats in their wheat and barley fields. The oats have responded genetically by producing semishattering and even nonshattering races. These hold most of their seed until harvest time and the cultivators reap the oats along with their primary crops. No effort is made to clean out the oat contaminants, so a mixture of seed is sown at planting time. Here, a species has automatically developed some of the genetic traits of a domesticated plant without any deliberate selection by man. It is a case of domestication by indifference.

Additional Examples

It is interesting to note that some crops may be partially domesticated and then abandoned and allowed to return to their wild state. Callen (1967) reports archaeological evidence for the domestication of a foxtail (*Setaria*) in Mexico. Setaria seeds show up in archaeological sites in some quantity, and it appears that over a period of time they gradually increased in size as though being selected under cultivation. Later they were abandoned when maize became available and replaced *Setaria* as a crop.

Li (1969) gives an account of vegetables in ancient China. Eight of the 32 discussed have now degenerated to the status of weeds. The most important green vegetable of ancient China, *Malva sylvestris* L., has become completely forgotten and has been relegated to the status of a weed. Its place was taken by *Brassica rapa* L., then a vegetable considered of only secondary importance. The most important tuber vegetable of ancient China was *Brassica rapa* L., though still commonly used, has been replaced by *Raphanus raphanistrum* subsp. *sativus* (L.) Domin, a relatively unimportant plant in China in former times.

Jacques Barrau (1965) pointed out that some of the earliest domesticated plants in the South Pacific have been almost abandoned. One of these, the Ti [*Cordyline fruticosa* (L.) A. Chev.], a plant used for religious and magic purposes and a symbol of clan permanence, was once eaten. It has been replaced in the diet by more desirable plants but lingers on in gardens for its magical and religious uses. In New Guinea, the tuberous plant *Pueraria montana* var. *lobate* (Wild.) Maesen & S.M. Almeida et Sanjappa & Predeep has been largely displaced by the sweet potato [*Ipomoea batatas* (L.) Lam.] and the name formerly given to *Pueraria* has been transferred to the sweet potato in relatively recent times.

Throughout the tropics the differences between wild, tolerated, encouraged, and cultivated are much less clear than in temperate zones. Plants are transplanted from the wild, brought into the garden, and escape again into the naturalized state where they are sometimes still harvested. The movement of useful plants from the wild condition to the cultivated and back again is a relatively simple and common occurrence.

The situation is by no means confined to the tropics, however. Weedy escapes from cultivation in temperate zones include spontaneous races of radish, carrot, lettuce, sunflower, false flax (*Camelina*), oats, rye, vetch, and many others. Furthermore, one man's weed is another man's crop. The wild oat may be a serious pest to the California wheat grower, but to the cattleman of the coast and foothill ranges, it may be the most important forage. Johnsongrass may be a hated weed to the Texas cotton farmer, but a valuable hay crop to his neighbor. The weedy watermelon [*Citrullus lanatus* (Thunb.) Matsum. & Nakai] is an obnoxious weed in cultivated fields over much of tropical and subtropical Africa, but in the dry season it may be the only supply of water for man and beast alike in the Kalahari Desert. *Brassica rapa* L. subsp. *oleifera* (DC.) Metzg. is sufficiently abundant to be damaging to wheat production in parts of the Andean Highlands, but is harvested for livestock feed and pot-herb greens by the native South American cultivators. A man may fight bermudagrass [*Cynodon dactylon* (L.) Pers.] with a passion in one field, deliberately plant an improved variety in another field, and nurture still another variety with tender care about his house.

A Short List of Cultivated Plants

For all the difficulties of definition, we can at least identify most of the important species that are cultivated and that usually have domesticated races. The origins are somewhat inconsistent, because as previously indicated, crops do not necessarily originate in definable centers. A number of those identified with the Near Eastern complex, for example, actually were domesticated in the Mediterranean Basin, northern Europe, or even western Europe. Crops that are added late to a complex, well-removed from the center of origin, are sometimes called "addition crops." Our information lacks precision in all but a few cases that have been intensively studied, and may well be faulty in other cases. The lists (Tables 3.1 and 3.2) will serve as a basis for further discussion.

There are, of course, hundreds more. Listed are only a few forage crops and plants producing rubbers, gums, resins, essential oils, dyes, poisons, tannins, ornamentals, or useful woods are not included. The lists of fruits, vegetables, and spices are far from complete and many familiar items are omitted. The tables do include the most important crops that provide food for the human species and is sufficiently comprehensive that some generalizations can be made.

First, it seems evident that man has searched out the plant kingdom rather thoroughly. The plants listed belong to about 55 families. Although most families contribute very little (e.g., Orchidaceae provides vanilla, Tropaeolaceae presents us with the *Tropaeolum*, Passifloraceae with *Passiflora*), an enormous percentage of the food for humankind is supplied by the Leguminoseae and Gramineae. Considering food plants only and discounting forages, drugs, narcotics, fibers, and so on, the grass family contributes 29 cereals plus sugarcane to the list and the legume family contributes 41 crops, mostly pulses, tubers, and edible pods and seeds. Other strong contributors are Solanaceae with 18 crops (fruits, spices, one tuber); Cruciferae, 13 crops (leafy vegetables, oil, root crops); Cucurbitaceae, 13 crops (squash, pumpkin, fruits, oil seeds); Rosaceae, 11 crops (mostly fruits); Liliaceae, 11 crops (edible bulbs); Umbelliferae, 9 crops (mostly spices and salad vegetables); and Araceae, 8 crops (all tubers).

Another conspicuous feature is the large number of vicarious domestications. If one species proves suitable for domestication, then a similar related species is likely to be useful as well. There are 40 genera on the list in which two or more species were domesticated independently. Some of the more

Table 3-1 A short list of cultivated plants and their probable origins in the Near East, Africa, Asia and the Pacific.*

The Near Eastern Complex		
Cereals		
Avena sativa L.	Oats; secondary crop	N. Europe
Avena strigosa Schreb.	Fodder oats; addition crop	Mediterranean
Hordeum vulgare L.	Barley; primary crop	Near East
Secale cereale L.	Rye; secondary crop	Anatolian plateau-N. Europe
Triticum aestivum L.	Bread wheat; addition crop	Transcaucasia-Caspian
Triticum dicoccum Schrank ex Schübl.	Emmer; primary crop	Near East
Triticum monococcum L.	Einkorn; primary crop	Turkey
Triticum timopheevii (Zhuk.) Zhuk.	Wheat (very minor)	Soviet Georgia
Triticum turgidum L.	Tetraploid wheat (derived from emmer)	Near East
Pulses		
Cicer arietinum L.	Chickpea; primary crop	Near East
Lathyrus sativus L.	Grasspea	Near East
Lens esculenta Moench	Lentil; primary crop	Near East
Lupinus albus L.	Lupine	Near East
Lathyrus oleraceus Lam. (syn.: *Pisum sativum* L.)	Garden pea; primary crop	Mediterranean
Vicia ervilia (L.) Willd.	Bittervetch	Near East
Vicia faba L.	Broadbean, fava	Wild form not known
Root and Tuber Crops		
Beta vulgaris L.	Beet, mangel, chard	Mediterranean, W. Europe
Brassica rapa L.	Turnip	Mediterranean (also maybe China)
Daucus carota L.	Carrot	Mediterranean, widespread
Raphanus raphanistrum subsp *sativus* (L.) Domin	Radish	Wild and weed races widespread

(*Continued*)

Table 3.1 (Continued)

The Near Eastern Complex

Oil Crops

Brassica napus L.	Rapeseed	E. Mediterranean
Brassica juncea (L.) Czern. (syn.: *Brassica nigra* (L.) Koch)	Mustard, mustard oil	E. Mediterranean
Carthamus tinctorius L.	Safflower	Near East
Linum usitatissimum L.	Flax, linseed	Near East
Olea europaea L.	Olive	Mediterranean
Papaver somniferum L.	Poppy	Near East

Fruits and Nuts

Corylus ssp.	Hazelnut, filbert	Balkans to Caspian
Cucumis melo L.	Melon	Near East
Cydonia oblonga Mill.	Quince	Balkans to Caspian
Ficus carica L.	Fig	Turkey, Iraq, Iran
Juglans regia L.	English walnut	Balkans to Pakistan
Phoenix dactylifera L.	Date palm	Lowland steppes of Near East
Pistacia vera L.	Pistachio	Turkey-Iran
Prunus amygdalus Batsch	Almond	Turkey to Pakistan
Prunus armeniaca L.	Apricot	Turkey-Iran
Prunus avium L.	Cherry	Balkans to Caspian
Prunus domestica L.	Plum	Balkans to E. Europe
Punica granatum L.	Pomegranate	Transcaucasia-Caspian
Pyrus communis L.	Pear	Turkey-Iran
Malus domestica (Sockow) Brokh. (syn.: *Pyrus malus* L.)	Apple	Balkans-Transcaucasia-Caspian
Vitis vinifera L.	Grape	Mediterranean

Vegetables and Spices

Allium cepa L.	Onion	Mediterranean
Allium sativum L.	Garlic	Mediterranean
Allium ampeloprasum L.	Leek	E. Mediterranean
Anethum graveolens L.	Dill	Mediterranean
Brassica oleracea L.	Cabbage, cauliflower, Brussels sprouts, kale, kohlrabi, broccoli	W. Europe

Table 3.1 (Continued)

The Near Eastern Complex		
Carum carvi L.	Caraway	Near East
Coriandrum sativum L.	Coriander	Near East
Cucumis sativus L.	Cucumber	India
Cuminum cyminum L.	Cumin	Near East
Foeniculum vulgare Mill.	Fennel	Mediterranean (widespread)
Lactuca sativa L.	Lettuce	Mediterranean
Lepidium sativum L.	Gardencress	Mediterranean
Petroselinum cispum (Mill.) Fuss	Parsley	Mediterranean
Pimpinella anisum L.	Anise	Mediterranean
Portulaca oleracea L.	Purselane	Mediterranean
Trigonella foenum-graecum L.	Fenugreek	Turkey
Fiber Plants		
Cannabis sativa L.	Hemp; widespread	Eurasian
Linum usitatissimum L.	Flax; primary crop	Near East
Starch and Sugar Plants (not root)		
Ceratonia siliqua L.	Carob, tree with sweet pods	E. Mediterranean
Forage Crops		
Agropyron spp.	Wheatgrasses	Eurasian, useful types from Turkey and European Russia
Agrostis spp.	Bentgrasses	W. Europe
Bromus inermis Leyss.	Smooth bromegrass	Turkey to Central
Dactylis glomerata L.	Orchardgrass, cocksfoot	Europe, Mediterranean
Lolium arundinaceum (Schreb.) Darbysh.	Tall fescue	Europe, N. Africa, Near East
Lolium spp.	Ryegrasses	Europe-Mediterranean
Medicago sativa L.	Alfalfa	Central Asia, Turkey-Iran
Medicago spp.	Medic clovers	Mostly Mediterranean
Melilotus spp.	Sweet clovers	Widespread Europe and Near East
Onobrychis viciifolia Scop.	Sainfoin	Turkey

(*Continued*)

Table 3.1   (Continued)

The Near Eastern Complex		
Phalaris arundinacea L.	Reed canary grass	Widespread Europe
Phalaris aquatica L.	Hardinggrass	Mediterranean
Phleum pratense L.	Timothy	Widespread Europe
Sorghum halepense (L.) Pers.	Johnsongrass	Mediterranean, Near East
Trifolium spp.	True clovers	Europe, Near East
Vicia spp.	Vetches	Mediterranean
Drugs, Narcotics, Fatigue Plants		
Atropa bella-donna L.	Belladonna	Mediterranean
Digitalis purpurea L.	Digitalis	Europe
Glycyrrhiza glabra L.	Licorice	Mediterranean, Near East
Hyoscyamus muticus L.	Henbane	Mediterranean, Near East
Papaver somniferum L.	Codeine, morphine, opium	Mediterranean
Plantago indica L.	Psyllium	Mediterranean

Africa		
Cereals		
Avena abyssinica Hochst.	Ethiopian oats	Ethiopia (from *A. barbata*)
Urochloa deflexa (Schumach.) H. Scholtz	Guinea millet	Guinea highlands
Digitaria exilis (Kippist) Stapf	Fonio	W. Africa – Nigeria to Senegal
Digitaria iburua Stapf	Black fonio	Nigeria to Togo, savanna
Eleusine coracana (L.) Gaertn.	Finger millet	Ethiopia-Uganda highlands
Eragrostis tef (Zuccagni) Trotter	Tef	Ethiopia
Oryza glaberrima Steud.	African rice	W. African savanna
Cenchrus americanus (L.) Morrone (syn.: *Pennisetum glaucum* (L.) R.Br.)	Pearl millet	Dry savanna Sudan to Senegal
Sorghum bicolor (L.) Moench	Sorghum	Savanna zones Sudan to Chad

Table 3.1 (Continued)

Africa		
Pulses		
Macrotyloma geocarpum (Harms) Mareechal & Baudet	Kersting's groundnut	W. African savanna
Lablab purpureus (L.) Sweet	Hyacinth bean	E. African savanna
Vigna unguiculata (L.) Walp.	Cowpea	W. Africa, forest margins
Vigna subterranea (L.) Verdc.	Bambara groundnut	W. African savanna
Root and Tuber Crops		
Dioscorea cayenensis Lam.	Yam	Ivory Coast to Cameroon
Dioscorea spp.	Lesser yams	Guinea to Cameroon
Plectranthus esculentus N.E. Br.	Kafir potato	W. Africa
Sphenostylis stenocarpa (Hochst ex A. Rich.) Harms	Yampea	W. Africa, forest zone
Plectranthus rotundifolius (Poir.) Spreng.	Piasa	W. Africa, becoming rare
Oil Crops		
Vitellaria paradoxa C. F. Gartn.	Karité, Butter tree	W. Africa, savanna
Elaeis guineensis Jacq.	Oil palm	W. Africa, forest margins
Guizotia abyssinica (L.f.) Cass.	Noog	Ethiopia, highlands
Ricinus communis L.	Castor bean, castor oil	E. Africa
Telfairia occidentalis Hook.f.	A gourd; oil from seeds	W. Africa
Fruit and Nuts		
Adansonia digitata L.	Baobab	African savannas
Blighia sapida K. D. Koenig	Akee apple, aril (toxic)	W. Africa
Citrullus lanatus (Thunb.) Matsum & Nakai	Watermelon	Dry savanna, S. and E. Africa

(Continued)

Table 3.1 (Continued)

Africa		
Vegetables and Spices		
Abelmoschus esculentus (L.) Moench	Gumbo, Okra	W. Africa
Aframomum melegueta K. Schum.	Malaguette	W. Africa, Ethiopa
Sesamum sesamoides (Endl.) Byng & Christenh.	Leaves and seeds	Savanna
Corchorus olitorius L.	Leaves and seedlings	Widespread
Melothria spaerocarpa (Cogn.) H. Schaef. & S. S. Renner	Leaves and fruits	W. Africa
Hibiscus sabdariffa L.	Calices and leaves	Widespread in savanna
Hibiscus cannabinus L.	Leaves and seeds	W. Africa
Piper guineense Schumach. et Thonn.	Seeds	W. Africa, forest
Sesamum alatum Thonn.	Leaves	Savanna
Sesamum radiatum Thonn. ex Hornem.	Leaves	Savanna
Solanum aethiopicum L.	Fruits	Savanna
Solanum macrocarpon L.	Leaves and fruits	Savanna and forest
Solanum spp.	Several "garden eggs" used for fruits and leaves	Sub-Sahara Africa
Fiber Plants		
Adansonia digitata L.	Baobab (bark)	Savannas, widespread
Gossypium herbaceum L.	Old World cotton	Sudan?
Hibiscus cannabinus L.	Kenaf	W. Sudan
Starch and Sugar Plants		
Ensete ventricosum (Welw.) Cheesman	Enset	Ethiopia
Parkia biglobosa (Jacq.) R. Br. ex G. Don	Tree with sweet pods	W. Africa, savanna
Sorghum bicolor (L.) Moench.	Sweet sorghum, sorgo	Savannas
Tamarindus indica L.	Tree with sweet pods	Tropical Africa

Table 3.1 (Continued)

Africa		

Forage Crops

Chloris gayana Kunth	Rhodes grass	Kenya to S. Africa
Cynodon aethiopicus Clayton et J. R. Harlan	A stargrass	Ethiopia to S. Africa
Cynodon dactylon (L.) Pers.	Bermuda grasses	Widespread
Cynodon nlemfuensis eriantha Steud	A stargrass	Kenya to S. Africa
Digitaria eriantha Steud.	Pangola grass	S. Africa
Eragrostis curvula (Schrad.) Nees	Weeping lovegrass	Tanzania to S. Africa
Eragrostis lehmanniana Nees	Lehmann's lovegrass	S. Africa
Hyparrhenia rufa (Nees.) Stapf	Jaragua grass	E. Africa
Megathyrus maximus (Jacq.) B. K. Simon & S. W. L. Jacobs	Guinea grass	Center in Kenya-Tanzania
Cenchrus clandestinus (Hochst. ex Chiov.) Morrone	Kikuyo grass	Kenya-Uganda
Cenchrus purpureus (Schumach.) Morrone	Elephant grass	Widespread, high rainfall
Sorghum bicolor (L.) Moench	Sorghum	Savanna zones

Drugs, Narcotics, Fatigue Plants

Coffea arabica L.	Coffee	Ethiopia, forest
Coffea canephora Pierre ex A. Froehner	Robusta coffee	Lowland forests
Coffea spp.	A few minor species	Forest zones
Catha edulis (Vahl) Endl.	Chat; leaves chewed	Ethiopia
Cola acuminata (P. Beauv.) Schott et Endl.	Cola; fruits eaten	W. Africa
Cola nitida (Vent.) Schott et Endl.	Chewed for caffein	W. Africa
Strychnos spp.	Trees; nux vomica and other uses	Widespread

(Continued)

Table 3.1 (Continued)

Africa		
Utility		
Lagenaria siceraria (Molina) Standl.	Bottle gourd	Widespread, origin unknown

The Chinese Region		
Cereals and Pseudocereals		
Echinochloa frumentacea Link.	Japanese millet	E. China
Fagopyrum esculentum Moench	Buckwheat	W. China
Fagopyrum tataricum (L.) Gaertn.	Tartar buckwheat	W. China
Oryza sativa L.	Rice	S. China to India
Panicum miliaceum L.	Proso, broomcorn millet	N. China
Setaria italica (L.) Beauv.	Foxtail millet	N. China
Pulses		
Glycine max (L.) Merill	Soybean	N.E. China
Mucuna pruriens (L.) DC.	Velvet bean	S. China
Vigna angularis (Willd.) Ohwi	Adzuki beans	S. China
Root and Tuber Crops		
Brassica rapa L.	Turnip	N. China (Mediterranean?; possible independent domestication)
Dioscorea esculenta (Lour.) Burkill	Chinese yam	S. China
Lilium tigrinum Ker Gawl.	Tiger lily (luxury item)	Temperate China
Nelumbium speciosum Thunb.	Lotus; seeds and tubers eaten	Widespread India and China
Raphanus raphanistrum subsp. *sativus* (L.) Domin.	Chinese radish, very large	Likely introduction from Europe
Sagittaria sagittifolia L.	An elephant ear	S. China

Table 3.1 (Continued)

The Chinese Region		
Eleocharis duleis (Burm.f.) Trin. ex Hensch	Chinese water "chestnut"	S. China
Oil Crops		
Vernicia fordii Hemsl. Airy Shaw	Tung oil	S. China
Brassica rapa L.	Rapeseed	Temperate China
Brassica juncea (L.) Czern.	A mustard seed oil	Temperate China
Sapium sebiferum (L.) Roxb.	Chinaberry tree	S. China
Fruits and Nuts		
Canarium album Blanco	Chinese "olive"	S. China
Carya spp.	Chinese hickories	Temperate China
Castanea henryi (Skan) Rehder. E. H. Wilson	Chinese chestnut	Temperate China
Chaenomeles spp.	Chinese quinces	Temperate China
Corylus spp.	Chinese hazelnuts	Temperate China
Diospyros kaki L.f.	Oriental persimmon	China
Pyrus loquat M. F. Fay & Christenh.	Loquat	Mountains of S.W. China
Ginkgo biloba L.	Ginkgo	N. China; known only in cultivation
Juglans regia L.	Walnut	Mountains of S.W. China
Litchi chinensis Sonn.	Litchi	S. China
Prunus armeniaca L.	Apricot	West temperate China
Prunus persica (L.) Batsch Stokes	Peach	West temperate China
Pyrus spp.	Chinese pears	Temperate China
Ziziphus jujube Mill.	Chinese jujube	West temperate China
Vegetables and Spices		
Allium chinense G. Don	Chinese shallot	Temperate China
Allium tuberosum Rottler ex Spreng	Chinese leek	Temperate China
Aralia cordata Thunb.	Udo	Eastern China

(Continued)

Table 3.1 (Continued)

The Chinese Region		
Benincasa hispida (Thunb.) Cogn.	Winter melon or wax gourd	Widespread
Brassica juncea (L.) Czern	Leafy vegetable	Temperate China
Brassica rapa L.	Chinese cabbage	Widespread temperate China
Cinnamomum burmanni (Nees & T. Nees) Blume	Spice	S. China
Cucumis melo L. var. *conomon*	Pickling melon	Widespread
Cucumis sativus L.	Cucumber	Widespread (probably India as well)
Lagenaria siceraria (Molina) Standl.	Bottle gourd; eaten young	Pantropical
Malva erticillata L.	The main leafy vegetable of ancient China; now a weed	China
Oenanthe javanica (Blume) DC.	Oriental "celery"	Wetlands
Stachys affinis Bunge	Chinese "artichoke"	Widespread
Wasabia japonica Matsum	Horseradish	Widespread
Zanthoxylum bungei Planch.	Chinese "pepper"	S. China
Zingiber officinale Rose.	Ginger	S. China
Zizania latifolia (Griseb.) Hance ex F. Muell.	A wild rice	Temperate China
Fiber Plants		
Abutilon theophrasti Medik.	Abutilon hemp	S. China
Boehmeria nivea (L.) Gaudich	Ramie	S. China
Cannabis sativa L.	Hemp	Central Asia
Drugs, Narcotics, Fatigue Plants		
Panax quinquefolius L.	Ginseng	Widespread
Arctium lappa L.	Burdock	Temperate China

Table 3.1 (Continued)

The Chinese Region		
Camellia sinensis (L.) Kuntze	Tea	S. and S.W. China
Cinnamomum camphora (L.) J. Presl.	Camphor tree	S. China
Rheum palmatum L.	Medicinal rhubarb	Temperate China
Utility		
Arundinaria spp.	Bamboos	S. China
Bambusa spp.	Bamboos; matting, houses, paper, pipes, etc.	S. China to Indonesia
Phyllostachys spp.	Bamboos; some used as food as well	China
Toxicodendron vernicifluum (Stokes) F. A. Barkley	Lac tree; varnish	S. China
Strobilanthes cusia (Nees) Kuntze	An indigo dye plant	S. China

Southeast Asia and Pacific Islands		
Cereals		
Coix lachryma-jobi L.	Job's tears, adlay	Indochina-Philippines
Digitaria cruciata Nees ex Hoof. f.	A millet	Hills of N.E. India
Oryza sativa L.	Rice	E. India to S. China
Panicum antidotale Retz. Lam.	Slender millet	Himalayas-Upper Burma
Paspalum scrobiculatum L.	A millet	Nilgiris region of S. India
Pulses		
Cajanus cajan (L.) Millsp.	Pigeonpea	India
Canavalia gladiata (Jacq.) DC.	A jackbean	S.E. Asia
Cyamopsis tetragonoloba (L.) DC.	Guar	India
Vigna unuculata (L.) Walp	A hyacinth bean	S.E. Asia

(*Continued*)

Table 3.1 (Continued)

Southeast Asia and Pacific Islands		
Psophocarpus tetragonolobus (L.) Taub.	Winged bean	New Guinea
Vigna aconitifolia (Jacq.) Marechal	Mat bean	S.E. Asia
Vigna umbellata (Thunb.) Ohwi & H. Ohashi	Rice bean	S.E. Asia
Vigna mungo (L.) Hepper	Urd, black gram	India or S. China
Vigna radiata (L.) Wilczek	Mung bean	India or S. China
Root and Tuber Crops		
Alocasia macrorrhizos (L.) G. Don	An elephant-ear	Indonesia-Melanesia
Amorphophallus spp.	Aroid tuber	S.E. Asia
Colocasia esculenta (L.) Schott.	Taro	Assam-Upper Burma
Cyrtosperma merkusii (Hassk.) Schott	An elephant-ear	Melanesia-Polynesia
Dioscorea alata L.	Winged yam	S.E. Asia
Dioscorea spp.	Several minor yams	S.E. Asia to Melanesia
Pueraria montana var. *lobata* (Willd.) Maesen & S. M. Almeida ex Sanjappa & Predeep	A yam-bean	Indonesia-Melanesia
Curcuma caulina J. Graham	Arrowroot	S. Pacific Islands
Oil Crops		
Brassica juncea (L.) Czern	Sarson	N. India
Cocos nucifera L.	Coconut	S. Pacific Islands
Sesamum indicum L.	Sesame	India
Fruits and Nuts		
Artocarpus altilis (Parkinson) Forstberg	Breadfruit	S.W. Pacific Islands
Artocarpus integer (Thunb.) Merr.	Jackfruit	S. Pacific and S.E. Asia
Averrhoa bilimbi L.	Bilimbi	S.E. Asia

Table 3.1 (Continued)

Southeast Asia and Pacific Islands		
Averrhoa carambola L.	Carambola	S.E. Asia
Citrus aurantiifolia (Christm.) Swingle	Lime	S.E. Asia and S. China
Citrus x *aurantium* L.	Sour orange	S.E. Asia and S. China
Citrus maxima (Burm.) Merr.	Shaddock, pomolo	S.E. Asia and S. China
Citrus x *limon* (L.) Osbeck	Lemon	S.E. Asia and S. China
Citrus medica L.	Citron	S.E. Asia and S. China
Citrus nobilis Lour.	Tangerine	S.E. Asia and S. China
Citrus x *paradisi* Macfad	Grapefruit; hybrid	West Indies
Citrus x *sinensis* (L.) Osbeck	Sweet orange	S.E. Asia and S. China
Durio zibethinus L.	Durian	S.E. Asia
Syzygium jambos L. (Alston)	Jambos, jambolans	S.E. Asia
Garcinia mangostana L.	Mangosteen	S.E. Asia
Mangifera indica L.	Mango	Indo-Malaysia
Musa acuminata Colla	Banana (A genome)	Malaysia-Thailand-Indonesia
Musa balbisiana Colla	Plantain (B genome)	Malaysia-Thailand-Indonesia
Musa sapientum L.	Banana (A – B hybrid)	E. India to Borneo
Musa spp. sect. Australomusa	Fe'i banana	Melanesia-Polynesia
Nephelium lappaceum L.	Rambutan	S.E. Asia
Dimocarpus longan Lour.	Longan	S.E. Asia
Vegetables and Spices		
Amaranthus spp.	Leaves and stems	S.E. Asia and possibly N. India
Curcuma longa L.	Turmeric	India-Malaysia
Elettaria cardamomum (L.) Maton	Cardamon	S.E. Asia

(*Continued*)

Table 3.1 (Continued)

Southeast Asia and Pacific Islands		
Syzygium aromaticum (L.) Merr. et Perry	Clove	Spice Islands
Myristica fragrans Houtt.	Nutmeg	Spice Islands
Piper nigrum L.	Black pepper	S.E. Asia
Solanum melongena L.	Eggplant	India
Fiber Plants		
Cocos nucifera L.	Coir fiber	S.W. Pacific Islands
Corchorus capsularis L.	Jute	India to Burma
Crotalaria juncea L.	Sun hemp	India-Burma
Hibiscus cannabinus L.	Kenaf	Originally from W. Sudan
Musa textilis Née	Manila hemp	Philippine Islands
Starch and Sugar Plants (not roots)		
Arenga pinnata (Wurmb) Merr.	Sugar palm	S.E. Asia and S. Pacific
Borassus flabellifer L.	Palmyra palm	S.E. Asia
Metroxylon sagu Rottb.	Sago palm	S.W. Pacific Islands
Metroxylon spp.	Palm spp.	Melanesia
Saccharum officinarum L.	Sugarcane	New Guinea
Tamarindus indica L.	Tree with sweet pods	Indian savanna (originally from Africa?)
Drugs, Narcotics, Fatigue Plants		
Areca catechu L.	Betel nut	S.E. Asia
Senna alexandria Mill.	Senna	S.E. Asia
Croton tiglium L.	Croton oil	S.E. Asia
Lawsonia inermis L.	Henna	S.E. Asia
Piper betle L.	Betel leaf; chewed	E. Asia
Piper methysticum Forst.	Kava-kava	Melanesia-Polynesia

* Species names were obtained from Plants of the World online (plantsoftheworldonline. org) by the Royal Botanic Gardens, Kew (accessed January, 2021). European origins were included in the Near Eastern Complex.

Table 3.2 A short list of cultivated plants and their probable origins in the New World.*

Mesoamerica & North America	South America
Cereals	
Panicum hirticaule J. Presl.; Sauwi	*Bromus mango* E. Desv.
Zea mays L.; Corn	
Pseudocereals	
Amaranthus cruentus L.; amaranth	
Amaranthus caudatus L.; amaranth, huauhtli	*Amaranthus caudatus* L.; amaranth, achis
Chenopodium berlandieri Moq; (more used as a vegetable); huaozontle	*Chenopodium pallidicaule* Aellen; cañahua
	Chenopodium quinoa Willd.; quinoa
Hyptis suaveolens L. Poit.; chia grande	
Iva annua L.	
Polygonum erectum L.; erect knotweed	
Salvia hispanica L.; chia	
Pulses	
	Arachis hypogaea L.; peanut
Canavalia ensiformis (L.) DC.; jack bean	*Canavalia plagiosperma* Piper; jack bean
	Inga feuillei DC.; pacae
	Lupinus mutabilis Sweet; chocho
Phaseolus acutifolius A. Gray; tepary bean	
Phaseolus coccineus L.; scarlet runner bean	
Phaseolus lunatus L.; lima bean	*Phaseolus lunatus* L.; lima bean
Phaseolus vulgaris L.; common bean	*Phaseolus vulgaris* L.; common bean
Root and Tuber Crops	
Bomarea edulis (Tussac) Herb.; sarsilla	*Arracacia xanthorrhiza* Bancr.; arracacha
	Calathea allouia Lindl.; leren
	Canna indica L.; achira

(*Continued*)

Table 3.2   (Continued)

Mesoamerica & North America	South America
Ipomoea batatas (L.) Lam.; sweet potato	*Ipomoea batatas* (L.) Lam.; sweet potato (poss.)
	Dioscorea trifida L.f.; yam
	Lepidium meyenii Walp.; maca
	Manihot esculenta; Crantz; manioc
	Maranta arundinacea L.; arrowroot
Pachyrhizus erosus (L.) Urb.; jícama	*Pacchyrhizus ahipa* (Wedd.) Parodi; jícama
	Pachyrhizus tuberosus (Lam.) Spreng.; jicama
	Oxalis tuberosa Molina; oca
	Polymnia sonchifolia Poepp; yacón
	Solanum tuberosum L.; potato
	Tropaeolum tuberosum Ruíz and Pavón; añu
	Ullucus tuberosus Caldas; ulluco
	Xanthosoma sagittifolium (L.) Schott.
Oil Crops	
	Arachis hypogaea L.; peanut
Helianthus annuus L.; sunflower	
Gossypium hirsutum L.; upland cotton	*Gossypium barbadense* L.; sea island cotton
Fruit and Nuts	
Manilkara zapota L. P. Royen; sapodilla	
	Anacardium occidentale L.; cashew
	Ananas comosus (L.) Merr.; pineapple
Annona macrophyllata Donn. Sm.; ilama	*Annona cherimola* Mill.; cherimoya
Annona glabra L.; anona	*Annona muricata* L.; guanábana
Annona purpurea Moc. et Sesse; anona morada	
Annona reticulata L.; anona	*Annona reticulata* L.; anona
Annona squamosa L.; sweet sop (poss.)	*Annona squamosa* L.; sweet sop
	Bertholletia excelsa Humb. & Bonpl.; Brazil nut

Table 3.2   (Continued)

Mesoamerica & North America	South America
Brosimum alicastrum Sw.; ramón	
	Bunchosia armeniaca (Cav.) DC.
Byrsonima crassifolia (L.) Kunth; nance	
Carica papaya L.; papaya	*Carica candicans* A. Gray; papaya
	Vasconcellea spp.; papayas
Casimiroa edulis La Llave & Lex; white sapote	
Casimiroa elulis La Llave; matasano	*Campomanesia lineatifololia* (Ruiz et Pav.); palillo
Crataegus mexicana D.C.; tejocote, Mexican hawthorn	
	Cyclanthera pedata (L.) Schrad.; achocha
	Solanum betacea Cav.; tree tomato
	Solanum circinatum Bohs; tree tomato
Diospyros ebenum J. Koenig ex. Retz.; black sapote	
Opuntia spp.; prickly pear	*Austrocylindropuntia subulata* (Muehlenpf.) Backeb.; cactus
Parmentiera aculeata (Kunth) Seem.; caujilote	
	Passiflora spp.; granadilla
Persea americana Mill; avocado	*Persea americana* Mill; avocado (poss.)
Persea schiedeana Nees; avocado	
Prunus serotina Ehrh.; capulín	
Psidium guajava L.; guava	*Psidium guajava* L.; guava
	Solanum muricatum Ait.; pepiño
	Solanum sessiliflorum Dunal.; coconá
	Solanum quitoense Lam.; lulo

(Continued)

Table 3.2 (Continued)

Mesoamerica & North America	South America
Spondias purpurea L.; jocote	*Spondias mombin* L.; cajazeira, hog plum

Vegetables and Spices

Capsicum annuum L.; pepper	*Capsicum baccatum* L.; pepper
	Capsicum chinense Jacq.; pepper
Capsicum frutescens L.; chili, aji	*Capsicum frutescens* L.; pepper
	Capsicum annuum L.; pepper, pimiento
Chenopodium berlandieri Moq.; huauzontle	*Cucurbita maxima* Duschesne.; squash
Cucurbita argyrosperma C. Huber; squash	*Cucurbita ficifolia* Bouché; squash
Cucurbita pepo (L.) Scheele.; squash, pumpkin	*Cucurbita moschata* Duchesne; squash
Lycopersicon esculentum Mill.; tomato (poss.)	*Lycopersicon esculentum* Mill.; tomato (poss.)
Physalis ixocarpa Brot. ex Hornem.; tomatillo	*Physalis peruviana* L.; uchuva
Sechium edule (Jacq.) Sw.; chayote	
Vanilla planifolia Jacks. ex Andrews; vanilla	

Fiber Plants

Agave spp.; agave	
Gossypium hirsutum L.; upland cotton	*Gossypium barbadense* L.; sea Island cotton

Forage Crops

Centrosema pubescens Benth.; centro (poss.)	*Centrosema pubescens* Benth.; centro
	Desmodium spp.; tick clover
	Stylosanthes gracilis Kunth.; stylo
	Tripsacum andersonii J.R. Gray
	Paspalum dilatatum Poir.; dallisgrass

Table 3.2 (Continued)

Mesoamerica & North America	South America
Drugs, Narcotics, Fatigue Plants	
Agave spp.; alcohol, agave	
Datura stramonium L.; jimson weed	*Datura* spp.; stramonium, Jimson weed
	Erythroxylun coca Lam.; cocaine, coca
Ilex vomitoria Aiton; yaupón	*Ilex paraguariensis* A. St.-Hil.; maté
Lophophora williamsii (Lem. ex Salm-Dyck) J.M. Coult.; peyote	
Nicotiana rustica L.; tobacco	
Nicotiana tabacum L.; tobacco	*Paullinia cupana* Kunth; guarana
	Paullinia yoco R. E. Schultes et Killip; yoco
Theobroma cacao L.; cacao, chocolate	
Utility	
Bixa orellana L.; achiote (poss.)	*Bixa orellana* L.; achiote, annatto
Crescentia cujete L.; tree gourd	*Crescentia cujete* L.; tree gourd (poss.)
Indigofera suffruticosa Mill.; añil	*Indigofera suffruticosa* Mill.; añil
Lagenaria siceraria (Molina) Standl.; bottle gourd	

* The table is arranged to point out the remarkable number of vicarious domestications in the Americas. It seems that if the native people of Mexico domesticated a species, natives of South America domesticated a similar species and vice versa. Uncertain domestications are indicated as possible (poss.). Species names were obtained from Plants of the World online (plantsoftheworldonline.org) by the Royal Botanic Gardens, Kew (accessed January, 2021).

frequently appearing genera are: *Annona*, 7 spp.; *Solanum*, 7 spp.; *Brassica*, 6 spp.; *Prunus*, 6 spp.; *Vigna*, 6 spp.; *Allium*, 5 spp.; *Capsicum*, 5 spp.; *Curcurbita*, 5 spp.; *Dioscorea*, 5 spp.; and *Phaseolus*, 4 spp.

The same kinds of plants were often selected in different parts of the world. Aroid tubers were domesticated in Asia, the South Pacific islands, South America, and possibly Africa (although not on the list). Yams were domesticated in Africa, Asia, and South America. Cotton was domesticated independently in Mexico, South America, and Africa or India, or both. In Tables 3.1 and 3.2, the genera with Old World–New World vicarious domestication are *Amaranthus, Canavalia, Dioscorea, Gossypium, Ipomoea, Lepidium, Lupinus, Prunus,* and *Solanum.*

Sometimes similar plants are put to quite different uses. In the Americas, amaranths were pseudocereals; in Asia they are pot-herbs. *Hibiscus cannabinus* L. and *Corchorus olitorius* L. are pot-herbs in Africa and fibers in India. *Lepidium* is a spicy salad green in the Near East but a root crop in the Andes.

Crops That Feed the World

While a great variety in food plants adds immeasurably to the quality of life, it is obvious that most of those listed in Tables 3.1 and 3.2 contribute relatively little to the nutrition of the world's population. Most of the food for humankind comes from a small number of crops and the total number is decreasing steadily. In the United States, in the past 70 years, many vegetables and fruits have disappeared from the diet, and the trend is going on worldwide. More and more people will be fed by fewer and fewer crops.

The major food crops have already been introduced, Table 2.1, Chapter 2, together with some information about them. Because they are so essential to human existence, more explanation is in order. The database was the FAO Production Yearbooks. The USDA Statistical Yearbooks also give figures for world production and production in selected countries. The two sources frequently do not agree, but are generally within the same order of magnitude. It is important to understand that all the figures are estimates and none is truly correct. Some estimates are better than others; one might suppose the more developed countries would have the most accurate figures, but even this idea could be suspect.

In any case, the gross production figures as reported in the yearbooks can be made more meaningful. Comparing apples with oranges is bad enough, but comparing grapes with wheat is absurd. Removing estimated moisture content and placing production on a dry matter basis is one improvement. However, rice is reported as paddy (rough rice in American terminology), which means the hulls are included. Coconuts, peanuts, and sunflowers are reported in the shell. We do not eat orange and banana peels or mango seeds, and so on. Removing the estimated wastages improves the estimates but does not solve all the problems.

A large proportion of maize, sorghum, barley, oats, rye, and soybean is fed to animals in the developed countries, but is human food elsewhere. Only part of the cottonseed oil is processed for human consumption, and a large fraction of grape production is used for wine. Reliable estimates for further corrections are hard to find.

Nevertheless, the overall pattern is so overwhelming that additional corrections would not change the basic situation. It is clear that the human species is currently an eater of grass seeds. We have become "canaries." It is also clear that the world's food supply depends on 12 or 15 plant species. It probably was not always so, although wheat, barley, rice, and maize have been the foundations of our high civilizations. The current trend is for the major crops to become even more major and for the lesser ones to dwindle. Efforts to change the pattern have had little effect to date.

References

Barrau, J. (1965). L'humide et al sec, an assay on ethnobotanical adaptation to contrastive environments in the Indo-Pacific area. *The Journal of the Polynesian Society, 74,* 329–346.

Callen, E. O. (1967). The first New World cereal. *American Antiquity, 32,* 535–538. https://doi.org/10.2307/2694082

Chase, A. K. (1989). Domestication and domiculture in northern Australia: A social perspective. In D. R. Harris & G. C. Hillman (Eds.), *Foraging and farming: The evolution of plant exploitation* (pp. 42–54). London: Unwin Hyman.

Dancette, C., & Poulain, J. F. (1968). Influence da l'Acacia albida sur les facteurs pédoclimatiques et les rendements des cultures. *Sols Africains, 13,* 197–239.

Harlan, J. R. (1989). Wild grass-seed harvesting in the Sahara and sub-Sahara of Africa. In D. R. Harris & G. C. Hillman (Eds.), *Foraging and farming: The evolution of plant exploitation* (pp. 79–98). London: Unwin Hyman.

Johnson, S. (1827). *Dictionary of the English language* (Vol. 3). London: Longman, Rees, Orme, Browne and Green.

Lévi-Strauss, C. (1950). The use of wild plants in tropical South America. In J. H. Steward (Ed.), *Handbook of South American Indians, Smithsonian*

Institution. Bureau of American Ethnology Bulletin 143 (Vol. 6, pp. 465–486). U.S. Government Printing Office.

Li, H.-L. (1969). The vegetables of ancient China. *Economic Botany, 23,* 253–260. https://doi.org/10.1007/BF02860457

Rindos, D. (1984). *The origins of agriculture: An evolutionary perspective.* Orlando, FL: Academic Press.

4

What Is a Weed?

. . . the history of weeds is the history of man.

Anderson (1954)

When you sow the berries of bays, weed not the borders for the first half year; for the weed giveth them shade.

Bacon (Johnson, 1827)

Source: Chapter title page sketches by Patricia J. Scullion, ASA/CSSA Headquarters Office.

Harlan's Crops and Man: People, Plants and Their Domestication, Third Edition.
H. Thomas Stalker, Marilyn L. Warburton, and Jack R. Harlan.
© 2021 American Society of Agronomy, Inc. and Crop Science Society of America, Inc.
Published 2021 by John Wiley & Sons, Inc.
doi:10.2135/harlancrops

Definitions

Because of the importance of weeds to agriculture and their probable roles in plant domestication, it is important that we have clearly in mind what is meant or implied by "weed." Some of the current definitions used in agronomic instruction, such as "a plant out of place" or "a weed is a plant that does more harm than good," are clearly inadequate. A weed is much more than that, but the implications of the term have changed over the years. The traditional use of the word is well expressed in the *Oxford English Dictionary* (Murray et al., 1961):

> Weed. 1. A herbaceous plant, not valued for use or beauty, growing wild and rank, and regarded as cumbering the ground or hindering the growth of superior vegetation.

In recent decades, however, it has become clear that identifying weeds by a value judgment is unsatisfactory. Biologists and laymen alike have become more ecologically minded than formerly. The terms "weedy," "weediness," or "weedishness" are commonly used and imply that a weed is a weed because of something it is or does and not simply because it is an object of prejudice. Bacon's line quoted by Samuel Johnson suggests that weeds should not be removed until they stop being useful and so weeds are not always unwanted. Bunting (1960) refers to "the weedy *Digitaria exilis*" as an important crop in parts of West Africa. Thus, if weediness implies unwantedness, then we are dealing with an unwanted crop, which is clearly nonsense.

In an essay on weeds, Harlan and deWet (1965) assembled a number of definitions of "weed," reproduced in Table 4.1. The professional weed men are all of the same mind and emphasize the unwanted qualities of weeds. This is understandable since their profession deals with control and eradication. Ecologists take a broader view of weeds. From this list, two basic themes to an understanding of weeds emerge: (a) a weed has certain characteristic ecological attributes, and (b) it is frequently unwanted because of these characteristics. It is the ecological behavior that is paramount. Human opinion has little to do with the ecological behavior of plants, but the ecological behavior of plants can have a lot to do with human opinion. Bunting (1960) put it this way:

> The common definition of a weed—that is a plant in the wrong place—conceals two important implications. Firstly, the word "wrong" implies a human opinion, since right and wrong are human concepts not inherent in nature. Secondly, the word "place" implies some characteristic dependence on environment, or in other words an ecological relationship, and clearly that relationship has to do with man's own botanical activities in farming.

Table 4.1 Definitions of weeds.

Source	Date	Definition
By professional weed experts		
Blatchley	1912	A plant out of place, or growing where it is not wanted
Georgia	1914	A plant that is growing where it is desired that something else shall grow
Robbins, Crafts, and Raynor	1942	These obnoxious plants are known as weeds
Fogg	1945	Any plant which grows where it is not wanted
Muenscher	1946	Those plants with harmful or objectionable habits or characteristics which grow where they are not wanted, usually in places where it is desired that something else should grow
Harper	1960	Higher plants which are a nuisance
Isely	1960	Any plant where it is not wanted, particularly where man is attempting to grow something else
Wodehouse	1960	An unwanted plant
Klingman	1961	A plant growing where it is not desired; or a plant out of place
Salisbury	1961	A plant growing where we do not want it
By enthusiastic amateurs		
Emerson (in Blatchley)	1912	A plant whose virtues have not yet been discovered
Cocannouer	1950	. . . This thing of considering all weeds bad is nonsensical!
King	1951	Weeds have always been condemned without a fair trial
By the ecologically minded		
Blatchley	1912	A plant which contests with man for the possession of the soil
Anderson	1954	Artifacts, camp followers
Rademacher	1948	Biologically speaking, weeds are plants that build up associations with useful plants and for which cultivation is beneficial or even necessary. Agriculturally speaking, weeds grow unwanted in cultivated land and there cause more harm than good. He then defines weeds in ecological terms as "pioneers of secondary succession"

(Continued)

Table 4.1 (Continued)

Source	Date	Definition
Dayton	1950	Introduced plant species which take possession of cultivated or fallow fields and pastures
Bunting	1960	Weeds are pioneers of secondary succession of which the weedy arable field is a special case
Isely	1960	The prime characteristic possessed by all important weeds is their ability to thrive in land subject to the plow
Pritchard	1960	Opportunistic species that follow human disturbance of the habitat
Salisbury	1961	The cosmopolitan character of many weeds is perhaps a tribute both to the ubiquity of man's modification of environmental conditions and his efficiency as an agent of dispersal

Source: Adapted from Harlan and deWet (1965).

Let us suppose a wheat farm in western Kansas is abandoned, not an unusual event over the last 100 years. For the first year or two after abandonment, the fields are covered with massive stands of sunflowers and Russian thistles (*Salsola kali* L., var. *tenuifolia* Tausch). These two species, the former native and the latter alien, are on everyone's weed list. But, the people have left and gone to town; there is nobody around to dislike them. Have they stopped being weeds? As a matter of fact, the weeds have now become useful plants in stabilizing the soil, preventing wind erosion, and reducing water erosion. It is true that weeds are often unwanted, but that is not what makes them weeds.

Harlan and deWet (1965) defined a weed as "a generally unwanted organism that thrives in habitats disturbed by man." Man has probably always caused some disturbance of habitats. Before he knew how to manipulate fire, man's disturbances were probably very minor and more or less limited to the vicinity of cave or camp. After he began to use fire to deliberately alter the vegetation, his disturbances were more widespread and more intense. Still, his set fires were relatively minor compared to the habitats he created after developing an effective agriculture system in which whole landscapes were churned up and entire floras destroyed and replaced by new vegetation.

The species adapted to the new, artificial habitats are mostly crops or weeds. Man generally wants the crops and tries by various means to encourage them; he does not want the weeds and tries by various means to eradicate them. Because both are adapted to the same habitats, however, practices that tend to favor crops also tend to favor weeds.

Since ecological behavior is the chief criterion for calling something a weed it would be logical to include animal species as well as plants in that category. The house sparrow, the starling, the statuary pigeon, the common brown sewer rat, the house mouse, the fruit fly (*Drosophila melanogaster* Meigen), and rabbits in Australia and New Zealand are excellent examples of animal weeds. Indeed, *Homo sapiens* is perhaps the weediest of all species, and the more he dominates the landscape, the more he seems to thrive. If we confine the concept of weeds to species adapted to human disturbance, then man is by definition the first and primary weed under whose influence all other weeds have evolved. One might argue that man is a domesticated animal rather than a weed. But man existed a very long time before he domesticated any other species; he has never seriously or consistently attempted to improve the race by selection or breeding as he has with other domesticates; and if we apply the test of unwantedness, the current alarm over the population explosion would appear to place man more in the category of weeds than domesticated animals. If man does succeed in controlling his own population size, we shall have an example of a weed becoming domesticated.

Intermediate States

There are two traditions with respect to weeds: one based on ecological behavior and one on man's response to the species in question. As might be expected of biological materials, neither criterion is sharp or clear-cut, and there are gradations between the extremes. With respect to ecological adaptation, the gradients might be diagramed in Table 4.1.

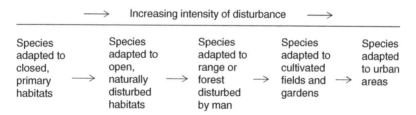

$\longrightarrow$	Increasing intensity of disturbance			$\longrightarrow$
Species adapted to closed, primary habitats $\longrightarrow$	Species adapted to open, naturally disturbed habitats $\longrightarrow$	Species adapted to range or forest disturbed by man $\longrightarrow$	Species adapted to cultivated fields and gardens $\longrightarrow$	Species adapted to urban areas

Figure 4.1 Species adaptation to environmental disturbances.

There were, of course, "disturbed" or unstable habitats long before man existed and they occur today in uninhabited regions. Natural disturbances of the kinds that would encourage pioneers of secondary succession or colonizers are common enough but usually do not affect very large areas. Examples would include river banks and frequently flooded areas; the shores of lakes, seas, and oceans; active dunes; areas unstable due to wind or water erosion; land slips; talus slopes; steep cliffs; land covered with volcanic ash or vacated by retreating glaciers; and so on. Species have evolved and adapted to all of these naturally disturbed situations.

There are also species adapted to disturbances caused by fires and blow-downs. Fires have always been a part of the natural environment of grasslands, woodlands, and dry forest and were so millions of years before man existed. Species have evolved that are resistant to fire and some even require occasional burning to survive. Some associations are so well adapted to periodic burning that man can cause as much disturbance of the habitat by controlling fires as he can be setting them.

Finally, animals other than man may cause wide-scale "disturbance," for example overgrazing by herbivores, traffic on game trails, trampling near water holes and bedding grounds, the rooting habits of some species, the burrowing habits of others, migrations in vast numbers of some gregarious species, and the work of termites, ants, and locusts.

Before agriculture, the most widespread disturbance was caused by the Pleistocene glaciation. Most of Europe and great sections of North America were alternately covered by ice and then exposed. Pioneer habitats were made available on a vast scale together with ample time for species to evolve that were adapted to such habitats. Thus, in temperate climates around the world, the chief weeds are Eurasian and North American species that developed in or near the areas of disturbance caused by Pleistocene glaciation.

The Pleistocene disturbance, however, was not nearly so vast or so rapid as that caused by man after the development of successful forms of agriculture. There is now hardly a spot anywhere on earth untouched by man in some way. Some weedy species might have been rather uncommon before man began to churn up the landscape, but when the agricultural "revolution" reached them their ecological niches were suddenly expanded and they prospered enormously as a result.

Most of our modern weeds and presumably all of our obligate weeds did not exist in their present form before agriculture. New products of evolution, they are dynamic and labile and constitute excellent subjects for the experimental study of evolution. They are products of vast disturbances on a continental scale where whole floras have been uprooted and replaced by imported vegetation and where masses of plants, separated for great periods of time, are suddenly brought together under conditions promoting mass hybridization. Such a global disturbance has probably never occurred before in so brief a time. The result has been the evolution of new plants adapted to the new ecological niches. These are what we call "weeds."

We have seen that whenever we deal with evolution we must deal with intermediates. Some species are weedier than others; some thrive under moderate disturbance but cannot tolerate intensive disturbance. Furthermore, some are disliked more than others. Plants with weedy tendencies may be encouraged as crops; others may be despised or hated. We could, therefore, diagram degrees of human response as we have done with disturbance of the habitat (Figure 4.2).

Edgar Anderson (1954) liked to think of weeds as plants that follow man around. Wherever man goes, he is soon surrounded by an array of plant companions whether or not he wants them, hates them, or ignores them. In Chapter 3, I mentioned *Acacia albida,* which moved out of its natural habitat along dry washes and spread over extensive areas of the African savanna. The seeds were mostly carried by livestock; the plants were spared by man because he thought them useful. Since the trees were not harvested, one can hardly call the white acacia a crop. The species does thrive under human disturbance, but the protection it received is deliberate and intentional. The karité and the oil palm have spread and thrived under somewhat similar conditions, but since they are harvested they can be called crops (Portéres, 1957).

At the other end of the scale are the really accomplished weeds that follow man in the face of hostility and outright warfare. Consider the dandelion (*Taraxacum*), for example. Millions of dollars and untold hours of toil are spent each year to reduce dandelion populations in lawns across the United States. Still they come, year after year, bespangling our green carpets with golden yellow blossoms and pushing up naked stems topped with seed heads of lacy gauze. To the unprejudiced eye, the flowers are really beautiful, but there is something in the culture that causes the owner of a lawn full of dandelions to feel guilty if not sinful.

Then there is crabgrass (*Digitaria sanguinalis*), so prolific a seed producer that it was once cultivated as a cereal in central Europe (Körnicke, 1885). Was it cultivated because it was so aggressive or did it become aggressive because it was once cultivated? We can detect no morphological changes in crabgrass that are typical of domesticated cereals.

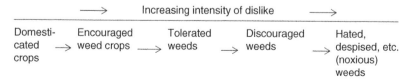

Figure 4.2   The spectrum of sentiment toward plants.

The seeds are small and shatter; European peasants used to cut it "half-ripe" to harvest the seed. In common crabgrass, cultivation did not lead to recognizable morphological changes, but it is possible that it led to more aggressive and competitive ecological races. The African crops, fonio (*D. exilis*) and black fonio (*D. iburua*), are more or less nonshattering and the seeds are considerably larger than those of wild or weed races of the same species.

The most noxious and despised weeds vary from region to region and crop to crop. It seems safe to state that there is no crop that does not suffer some yield decreases from competition with weedy plants, but weeds can occur in nonagricultural land as well. An example of an urban weed is *Ailanthus,* the tree of heaven. It has an astonishing affinity for brick and concrete and is often found growing in cracks in pavement or walls of buildings.

As pointed out in the last chapter, one man's weed is another man's crop and vice versa. Some crops undoubtedly originated from weed progenitors and some crops have degenerated into weed races. Many intermediate states exist. Plants drift in and out of cultivation, are domesticated, abandoned, ennobled, and may degenerate again; they escape, become naturalized, migrate, retreat, build hybrid swarms, and evolve new races. We shall examine a few cases of genetic interaction between wild, weed, and cultivated races.

Crop-Weed Complexes

The evolution of weeds often parallels the evolution of crops and the same principles apply to both. Both weeds and crops often begin with a common progenitor, as in those complexes where each crop has a companion weed. There are weed and cultivated races of einkorn, barley, sorghum, rice, oats, pearl millet, potato, tomato, pepper, sunflower, carrots, radish, lettuce, and many others. Perhaps most cultivated plants have one or more companion weed races. In some cases, the weed races can be easily distinguished from the wild forms; in others, it is extremely difficult.

The situation in sorghum is quite clear and it can be used as a general model for weed-crop complexes. Massive stands of truly wild races of sorghum can be found widely distributed over the savanna zones of Africa. They are often far removed from human disturbance, and represent truly wild grasses. The same materials can, however, be rather weedy when the habitat is disturbed. With the building of the Aswan Dam, it was necessary to move a rather large number of people out of the area to be flooded by Lake Nasser. People from Wadi Haifa in northern Sudan were moved to an area near Kassala, close to the Ethiopian border. This area was covered with vast stands of truly wild sorghums. The land was leveled, an irrigation project was established, ditches and drains were constructed, and soon irrigated farms covered many thousands of hectares of the Sudanese savanna. The wild sorghum has survived as a weed in the cotton and wheat fields and along the irrigation ditches of this project. There has been no evident genetic change in these populations, and morphologically the field weeds of the Kassala project are identical with the wild material nearby. The people from Wadi Haifa preferred to grow wheat rather than cultivated sorghum so that there has been little or no interaction between the wild sorghum and cultivated sorghum in that region. Although this race has done well as a weed, it is not so aggressive as to be very troublesome, and stands on the farm lands are not as dense as those in rangeland not far away.

In other parts of Africa, there is a conspicuous interaction between stands of wild sorghum and cultivated forms. The result is a race, often called shattercane, that is a very serious pest. This weed infests fields of cultivated plants on a massive scale and is extremely difficult to control. It is recognizable morphologically and easily distinguished from the wild races.

The fact that shattercane interacts with cultivated sorghum is clearly shown by the way it mimics the particular race of cultivated sorghum with which it is growing. Shattercane in Sudan and parts of Ethiopia where the durra race is the most common cultivated sorghum tends to have semicompact or even compact heads just like the cultivated kinds. The spikelets shatter and have a typical shattercane morphology. On the highlands of Ethiopia where the cultivated sorghum has a loose, open head, shattercane also has loose, open heads.

Wild sorghums shatter by means of a callus or abscission layer. In the course of domestication, callus formation was suppressed genetically. In most studies this suppression appeared to be controlled by a single recessive gene. Back mutation is extremely rare, but an alternative shattering device has evolved. Some shattercanes disseminate seed by breakage of inflorescence branches just below the point where the callus would have formed. Thus, the shattered spikelet pair carries a short branch fragment attached. None of the shattercanes that we have been able to examine as escapes in the United States develop calluses. In Africa, where wild species are commonly available for hybrid formation, shattering by callus formation is the mode of dispersal. The branch fragment is an excellent marker to help identify secondarily derived weed races. The evidence is good that weed races can evolve from cultivated races as well as from wild × cultivated hybridizations; and in sorghum, the two kinds of weed races can be identified morphologically.

Thus, there are four clear-cut and morphologically recognizable categories for sorghum: (a) truly wild races that can tolerate considerable disturbance of the habitat and are mildly weedy, (b) shattercanes that are derived from wild × cultivated crossing and are serious pests, (c) shattercanes that are derived secondarily from cultivated races and are also serious pests, and (d) semidomesticated to fully domesticated races that are grown under cultivation. The range of variation is not continuous, however. The shattercanes resemble domesticated races more closely than wild races.

Gene flow and hybridization between johnsongrass (*Sorghum halepense* (L.) Pers.) and cultivated sorghum has been established with molecular data (Morrell et al., 2005; Ohadi et al., 2017) and the aggressiveness of johnsongrass in the United States may be partly due to the infiltration of germplasm from cultivated sorghum. For one thing, we have not found the American type *S. halepense* anywhere except in the United States. The Mediterranean and Asian races of the species are quite different and are relatively unaggressive. The Mediterranean race is a small, slender, spindly type that grows slowly and is not much of a weed problem. The Indian race (*miliaceum*) is very tall, but does not behave at all like johnsongrass nor does it look much like it. We suspect the wild races become better adapted to field conditions through hybridization with cultivated races. Since johnsongrass is a tetraploid, the genetic infiltration was probably almost all in one direction from cultivated to weedy sorghum. Hybrids are not uncommon and a backcross to johnsongrass quickly suppresses characters of cultivated races. I did find, however, a colony of *S. halepense* in Mexico that shattered by branch breakages instead of by callus, suggesting it had cultivated sorghum in its background.

Extensive genetic interaction between wild, weed, and cultivated races can be demonstrated in African rice (*Oryza glaberrima* Steud.) in West Africa. The wild forms are easily distinguished from the weed races that infest the rice fields in that region. Until recently, most of the rice cultivated in West Africa was of the Asian kind (*O. sativa*), which had almost completely replaced the original native African rice. The African weed rice, however, has persisted better than the cultivated crop and is a serious pest in fields of Asian rice. In addition, the cultivated glaberrima rices were not all fully domesticated and some of them have escaped as weeds of rice fields as well. The populations that build up are sometimes extraordinarily complex. To add to the complexity are new interspecific rice cultivars becoming common in Africa. Due to the tenatious properties including drought, disease, insect, and weed tolerance of the African rice populations, the Africa Rice Center has used them to release new cultivars of *O. glaberrima* x *O. sativa* hybrids called NERICA (New Rice for Africa) rice.

In Asia, the races are not so clearly recognizable morphologically. Weed rices are serious pests of rice fields in India, Myanmar, Cambodia, Indonesia, Vietnam, and Laos Peninsula, and they frequently hybridize with cultivated types. A number of studies have been made that show that shattering types frequently segregate for characters of cultivated rice (Ghose et al., 1956; Mitra & Ganguli, 1932; Ramiah & Ghose, 1951). The agronomic difficulties were described clearly by Bhalerao (1928):

> Due to natural crossing with wild *Oryza* species, every cultivated variety has its own grain shedding type which cannot be distinguished till the last stage of panicle development when all the grains on the panicle shed down in the field below. In an attempt to solve the problem, some plant breeders developed a series of purple-leaved cultivars so that the farmers could weed their fields before maturity. This seemed like a good idea, for all the cultivators had to do was pull out the green plants and leave the purple ones. In a few years, however, a purple-leaved weed race had evolved and the farmers were no better off than before.

The genetic interactions among wild, weed, and cultivated races of barley are perhaps less common than those among rice in India, but hybrids and hybrid derivatives can be found. The most conspicuous are those involving wild or weed two-rowed barley and six-rowed cultivars. Brittle six-rowed types are produced, but they are poorly adapted and soon disappear. More lasting effects of introgression can be detected in seed colors, rough and smooth awns, and so on.

In Near Eastern barley, some races can be identified as wild, but in other races the wild and weed forms are confounded. There is a small wadi race that appears to be truly wild. Found in wadi bottoms from the Negev to Afghanistan (Harlan & Zohary, 1966), it is very small, slender, and grassy, and has small ears, small seeds, and short awns. It may often be found far removed from field agriculture. In more mesophytic races of barley, however, it is very difficult to distinguish the weed from the wild type. Roadsides, edges of fields, and waste places are often very densely populated with weed barley. However, the same kind of barley may be found in fairly primary habitats and there seems to be no good way to separate them morphologically. The only difference in this case is in their habitats. They are considered spontaneous forms, which means that both wild and weed forms seed themselves without deliberate planting by man.

Although the wild emmers of the Near East are not particularly weedy, they do come into contact with cultivated wheats sufficiently to cross occasionally. Genetic characteristics have been observed to move in both directions: from wild into cultivated and cultivated into wild (Zohary, 1971). In the wheats of the Near East, crosses sometimes take place between forms of different ploidy levels. Hybrids between 2x and 4x, 2x and 6x, and 4x and 6x have been observed. The triploids and pentaploids are not always completely sterile and backcrosses can restore fertility through the function of unreduced eggs (Zohary, 1971). These crosses frequently involve weedy species of *Aegilops* and can result in substantial increases in variation.

The races of wild and weed teosintes have been studied in some detail (Wilkes, 1967). The Chalco race, considered a weed because it grows in cultivated fields only, is said to have more maize characteristics than other races. Races from the Rio Balsas watershed may thrive on steep slopes without cultivation and are considered to be closer morphologically to wild types. Some races from Guatemala are even more extreme in showing a wild-type morphology and an adaptation to less disturbed habitats.

Some Weed Adaptations

An interesting adaptation syndrome is one in which the weed mimics the crop sufficiently well that the seed is harvested along with the crop and sown with it at the next planting season. *Camelina sativa,* subsp. *linico-la* (L.) Crantz is a well-known example; races have developed that resemble particular varieties of flax in stature, posture, and maturity as well as in seed size and weight. *Echinochloa crus-galli* var. *oryzicola* (L.) Beauv resembles the rice plant very closely throughout its development, from seedling to flowering time (Yabuno, 1961). This makes it very difficult for the cultivator to weed his rice fields in the early stages and at flowering time he is reluctant to walk through the rice fields to pull out the weed. Another example of a weed adaptation is a race of *Bromus secalinus* L. that retains its spike-like panicle intact at maturity, whereas most races of the species fragment and shatter their seeds. The nonshattering trait insures that the weed will be harvested with the cultivated rye with which it commonly grows. A race of weed rye that is semibrittle has some seeds that shatter and fall to the ground infesting the soil; the remainder are nonshattering and are harvested along with the wheat crop. These are then planted with the wheat seed the following season. Ethiopian oats have already been described as a crop that originated by the same mechanisms. It seems evident that these adaptations evolved as a result of manipulation by man and are not characteristic of wild plants.

The weed floras of mine dumps have been analyzed in recent decades. Some dumps have such high concentrations of zinc, lead, copper, and other heavy metals that they are very toxic to most plant life; only the most metal-tolerant genotypes of the most tolerant species can grow in such habitats. Genetic studies have revealed striking differences in tolerance among genotypes of the same species. The distribution of tolerant genotypes corresponds to depth and distribution of toxic concentrations of the poisonous metals (Antonovics, 1971). Heavy applications of salt to streets and highways in wintertime have resulted in strong selection for salt-tolerant genotypes of roadside plants in some regions. Weeds have responded to selection pressures imposed by various other pollutants of the industrial age illustrating, again, that they tend to be genetically labile and capable of rapid evolution.

Most weeds are characterized by enormous phenotypic plasticity. Under favorable conditions a given genotype will be tall, robust, well developed, and highly productive. Under unfavorable conditions the same genotype may be minute and depauperate, live only a short time, and produce few seeds. In a paper entitled "The Weedishness of Wild Oats," H.V. Harlan (1929) described the remarkable behavior of a wild oat population in a barley nursery. The nursery contained winter forms which were still in winter rosettes while spring forms in the same field were tall and heading out, with some of the earliest varieties maturing. In barley, these differences in growth habit were genetically controlled, but the wild oats with which the field was infested produced phenotypic mimics of all the growth habits. When grown with winter barley, the wild oats produced a low winter rosette; in adjacent rows of spring barley, the wild oats were tall and heading out. As the early barley was maturing, the wild oats were ripening. All stages could be seen on the same day. The capacity for phenotypic mimicry is under genetic control and constitutes an excellent adaptive mechanism for weeds.

Of course, there are many species of weeds that are neither closely related to cultivated plants nor do they mimic them. For many weeds, prodigious numbers of seeds are produced and these have special adaptations that prevent them from all germinating at the same time. The seeds are often small and capable of staying viable in the soil for long periods of time. This dormancy may be due to a variety of special adaptations and may be separated in various ways. Light-sensitive seeds buried during tillage may remain viable for a number of years; when the soil is turned again some are brought to the surface and sprout. Some seeds have a cold requirement, others have an inhibitor which breaks down over time or can be leached out, certain seeds are stimulated by specific chemicals, and some will not germinate except in the presence of certain other plants. Most annual weeds appear to have an adaptation syndrome involving the production of enormous numbers of seeds with special mechanisms insuring that they will not all sprout at once.

Among perennials, most of the adaptations for weediness concern longevity. Some have rhizomes that not only store food reserves for regrowth but also are easily distributed by tillage implements. Some have very deep taproots and many produce buds so deeply that they sprout from below the plow line. Other perennials are woody root-sprouters which are not killed by simply being cut down or burned. Some of the more objectionable weeds are protected by thorns, stinging hairs, or poisons in addition to being persistent. Whatever the adaptations may be, simple or elaborate, they tend to fit the weed to a particular niche of the human habitat, frequently with such success that they cost us dearly in control measures.

Weeds and History

Anderson's statement that the history of weeds is the history of man (Anderson, 1954) needs amplification. In his view, weeds follow man like fruit flies follow a ripe banana or a gourd of unpasteurized beer. Wherever man goes, he is surrounded by his weedy companions because he is a chronic disturber of habitats, and this was going on long before agriculture. If a family or band of hunters-gatherers lived in a cave or an open camp for any length of time, there would be the usual refuse heap, attrition of the local vegetation, and disturbance of the soil opening up a habitat for colonizers. In his dump-heap theory of domestication, Anderson visualized gatherers taking advantage of the dense and luxurious stands of weeds, which had themselves taken advantage of the open habitats and enriched soil of the refuse heap (Anderson, 1954). It is a sort of ecological propinquity theory in which man and plants share the same habitats that man himself has created.

It is true that some crops have a dump-heap look about them, for example, Chenopods, amaranths, sunflowers, cucurbits of various sorts, and so on. The eastern North American complex of *Iva, Chenopodium, Polygonum, Cucurbita, Ambrosia,* and *Helianthus annuus* is just such a weedy complex one would expect on dump-heaps (Fowler, 1971; Watson, 1989). On the other hand, swamp-dwelling plants such as taro (*Colocasia*), *Alocasia, Cyrtosperma, Trapa,* sago palm, and rice would appear to be poor candidates for such an origin. Wild wheat and wild sorghum occur in denser stands in near-climax formations than on dump heaps. The theory fits some species, but not others.

After agricultural systems had evolved, the role of weeds in human history became more conspicuous and easier to follow. Eloquent testimony is found in pollen profiles in many parts of the world. The pollen and spores of plants tend to be extremely durable and, under certain conditions, may be preserved as fossils in excellent conditions for millions of years. The pollen grains are frequently recognizable morphologically and can be identified as to genus or, occasionally, to species. A good pollen profile yielding a chronological sequence of pollen populations can tell us a great deal about past changes in vegetation. The discipline of pollen studies is called *palynology*.

Perhaps the best sources of pollen profiles are sediments from lakes and swamps. Wind-borne pollen from the surrounding vegetation is shed and falls on the water year after year. Some of the pollen sinks to the bottom where oxygen content is low and biological activity suppressed. It usually settles along with the clay and silt that reach the lake from upstream, and once buried, may be preserved for long periods of geologic time. Special coring devices have been developed to sample sediments of this kind. Cores must be extracted with care and precautions taken against contamination. Sometimes enough organic matter is encountered in a core profile that ^{14}C dates can be determined for various depths of sediment.

Pollen profiles can also be developed from soils, refuse heaps, and archaeological sites. In well-aerated soil, however, preservation is poor and the number of grains recovered may be small. However, palynological studies have produced information about changes in vegetation during the Pleistocene and on into the more recent past. The quality of information depends upon the volume of studies, and evidence from one or two fragmentary profiles can be misleading.

Palynology is well developed in Europe, and a large amount of information on postglacial vegetation is available. Forests spread over the land as glaciers retreated and successional changes in populations can be followed by the pollen sequences. In scattered areas, the profiles show a sudden decrease in tree pollen and a dramatic increase in the pollen of field weeds. Neolithic farmers had arrived. The companion weeds to crops give us the news. The early farmers of Europe practiced some sort of shifting cultivation, opening up the forest with the aid of fire, and in soil profiles there is sometimes evidence of much burning and alteration of the soil itself (Dimbleby, 1967; Halstead, 2017).

Pollen profiles have helped trace the spread of agriculture in the Near East, eastern North America, and a few other parts of the world. Unfortunately, palynology is not well developed in the tropics and our evidence from some of the most critical regions is meager and tenuous to say the least. Pollen cores from Taiwan suggest possible man-made disturbances of the forests as early as 12,000 BP, but the evidence becomes more substantial at about 6,200 BP (Chang, 1970). Changes in distribution and composition of African forests have been traced over time by the pollen record, but none of the changes observed through the 1980s could be directly attributed to agricultural activities (Hamilton, 1972; Livingstone, 1984). However, as agriculture has become more intense with population growth, the forest has been highly modified and is significantly smaller than in prehistoric times (Norris et al., 2010). As more palynological evidence accumulates we should be able to develop a much clearer picture of agricultural dispersals, and a large portion of the information will be provided by weedy plants.

On a more recent timescale, the present distribution of weeds may tell us about human activities of the past. Yarnell (1965) reported on a study of plants that occur on or near the Native American Pueblo ruins in New Mexico. Some species are common on the ruins, but they are infrequent, rare, or absent away from the site. Almost all of these plants are known to have been used by Native Americans for some purpose. A few were semicultivated food plants, but a number of them were medicinal or ritual plants, not normally cultivated but weedy enough to become established. Gilmore (1930), Moseley (1930), and others have pointed out cases in which the distribution of species in S.W. North America can best be explained by their dispersal by Native Americans.

Mayan Old Empire ruins can be spotted all over the Yucatan Peninsula by stands of ramon. The groves are locally called *ramonales,* and commonly include trees of mamey (*Calocarpum mammosum* (L.) Pierre), zapote (*Manilkara zapota* (L.) P. Royen), guayo (*Talisia olivaeformis* (H.B.K.) Radlk.), aguacate (English: Avacado) (*Persea americana* Miller), custard apple (*Annona reticulata* (L.) P. Royen), and black zapote (*Diospyros* spp.). Some of these are not found in the wild elsewhere in their region. To what extent these fruit trees were planted by the ancient Maya and to what extent the stands were built up by protection and encouragement, we do not know. In either case the groves were established artificially and have maintained themselves in some cases for 1000 years (Lundell, 1938).

Hutterer commented (1983, p. 176): "Today, the growth of (native) tobacco in rock shelters and in the mouth of caves in central Australia is usually a good sign that these sites were used prehistorically by aborigines." Akihama and Watabe (1970) made a detailed study of the distribution of weed rice in Thailand. They concluded that weed rices were most abundant in regions where rice cultivation was very old. The densest populations seemed to be associated with ancient tanks and irrigation systems. The weeds have a story to tell if only we knew enough to interpret the evidence properly.

Studies of the biosystematics of *Cynodon* led to an examination of historical dispersals. *Cynodon incompletus* Nees is a species endemic to South

Africa, but it is found at wool-cleaning works in Bedfordshire, Worcestershire, and Yorkshire in England, and in Filburg and Gaviers de la Vesdre in Belgium (cf. specimens at Kew). Wool-cleaning stations have developed their own peculiar mini-floras, including plants associated with sheep-rearing from various parts of the world.

Common bermudagrass (*Cynodon dactylon*) is widespread in the Old World, but has sorted out into geographical, ecological, and morphological races. Clones from central Europe, southern Europe, southeastern Europe, the Near East, Afghanistan, East Africa, South Africa, and the wet tropics are all different and most of them recognizable. Materials naturalized and escaped in the United States can often be traced with some assurance to their region of origin by appearance and growth habit (Harlan, 1970; Harlan & deWet, 1969; Harlan et al., 1970).

The earliest herbarium specimens and records of bermudagrass in the United States are from naval stations and ports along the East Coast. It had repeatedly been introduced with ship's ballast, and many other weeds also found their way to our shores the same way. By the early 19th century, bermudagrass was being extolled for pasture and levee stabilization from Georgia to Mississippi (Affleck, 1884; Moore, 1958; Spalding, 1844). Others were beginning to curse it as "wiregrass" and "devilgrass."

Bermudagrass is one weed that seems to have made its way to practically every island, large or small, in the Pacific Ocean. All of the early specimens we have seen are of the small, turfy race from the tropics. Specimens collected from some of the larger islands during the last 50 years indicate a different race that is diploid with very long, fast-growing stolons that is native to the drier zones from South Africa northward toward East Africa to Isreal and southeastward to India. In the city of Honolulu, the small tropical race is found throughout the older parts of town. Along the new freeways and in the newly developed suburbs, the diploid race has taken over, and some of the residents have bermudagrass stolons overtopping hedges 2 m high.

Weeds and history are closely associated in the Pacific area. Weedy escapes can help trace migrations of people over periods of time. The seedy fe'i bananas were introduced to Tahiti by Polynesians and became a naturalized element of the flora, and traces of ancient voyages can be detected throughout the Pacific (Barrau, 1963). The island of Guam has a large number of American weedy species that are attributed primarily to the Manila-Acapulco shipping that maintained a regularly scheduled route from 1565 to 1815; Guam was a refreshing station for ships sailing both directions. The land around Manila Bay also has a large number of weedy introductions from America (Merrill, 1922).

Sources of weeds can be traced if we know enough about them. Weeds do tell us something about where man has been and what he has done. Anderson was right; the history of weeds is the history of man, but it is an obscure history full of gaps and subject to misinterpretation because we have not taken the trouble to study the most common plants about us.

Weeds are not always passive companions to man; sometimes they alter human behavior. Alang-alang or cogongrass [*Imperata cylindrical* (L.) Beauv.] is an aggressive rhizomatous grass of the Old World tropics. It becomes readily established in fields cleared from the forest. If the fields are cultivated too many seasons in succession, cogongrass may take over completely and make it very difficult to grow crops. Furthermore, the forest tree seedlings and root spouts may be suppressed, making it almost impossible to employ the bush fallow of traditional shifting cultivation. Large areas of grassland, sometimes called "cogonals," may develop in regions of forest climax. With traditional agricultural techniques the cogonals are virtually worthless and people may have to move their villages elsewhere. Forest succession may be very slow in cogonals, even after abandonment by man. As a consequence, there is strong pressure to keep the cropping cycles short in order to avoid cogonal development.

Aggressive, rhizomatous weeds have always caused serious problems in agriculture. Most of the worst ones are grasses and include: cogongrass, kikuyugrass, wild sugar cane (*Saccharum spontaneum* L.), bermudagrass, and quackgrass (*Agropyron repens* (L.) P. Beauv.). All of these have from time to time forced man to abandon his fields. On the other hand, all but cogongrass and wild sugar cane can be useful and productive forage grasses.

Sometimes aggressiveness has its virtues. After the conquest of Peru, the Spaniards introduced European livestock that soon caused very severe problems associated with overgrazing. The South American grasses had not evolved under heavy grazing use and were easily obliterated. The denuded mountainsides suffered enormously from erosion and streams were clogged with sediments. In due time, bermudagrass and kikuyugrass found their way to Peru and have done great service in stabilizing slopes at lower and middle elevations.

The coast and foothill ranges of California provide a celebrated case in which the native grassland component of the flora has virtually been replaced or overwhelmed by an aggressive weed flora from the Mediterranean. The wild oats (*Avena barbata* and *A. fatua*), filarees (*Erodium*), mustards (*Brassica* spp., *B. juncea* (L.) Czern., *B. napus* L., *Sisymbrium officinale* L., *Sisymbrium altissimum* (L.) Scop.), bur-clover (*Medicago hispida* Gaertn.), weed radish (*Raphanus raphanistrum* subsp. *sativus*), Klamath weed (*Hypericum perforatum* L.), white clover, sweet clover, chickweeds, and others are among the most common and conspicuous. It will be noted that most of them are related to cultivated species and several have domesticated races. The California climate is well suited to the production of annuals and the coast and foothill ranges are more productive and stable on a sustained basis than they were before the weed flora was introduced.

Conclusions

Weeds are adapted to habitats disturbed by man. They may be useful in some respects and harmful in others. They may be useful to some people and hated and despised by others. There are weed races of most of our field crops and these interact genetically with cultivated races as well as truly wild races. This interaction probably results ultimately both in better crops and more persistent weeds. Some weeds have evolved elegant adaptations under the influence of man and many had weedy tendencies before man existed. Weeds are products of organic evolution; they exist in intermediate states and conditions. They are also genetically labile and phenotypically plastic. Weeds have been constant and intimate companions of man throughout his history and could tell us a lot more about man, where he has been and what he has done, if only we knew more about them.

References

Affleck, T. (1884). Letter to the editor. *American Agriculturist, 3*, 335–336.

Akihama, T., & Watabe, T. (1970). Geographical distribution and ecotypic differentiation of wild rice in Thailand. *Tonan Ajia Kenkyu, 8*, 337–346.

Anderson, E. (1954). *Plants, man and life*. London: A. Melrose.

Antonovics, J. (1971). The effects of heterogeneous environment on the genetics of natural populations. *American Scientist, 59*, 593–599.

Barrau, J. (1963). *Plants and the migrations of Pacific peoples: A symposium*. Honolulu, HI: Bishop Museum Press.

Bhalerao, S. G. (1928). The wild rice (*Oryza sativa*) of the Bombay Presidency. *Poona Agricultural College Magazine, 20*, 45–49.

Blatchley, W. S. (1912). *The Indiana weed book*. Indianapolis, IN: Nature Publishing Company.

Bunting, A. H. (1960). Some reflections on the ecology of weeds. In J. L. Harper (Ed.), *The biology of weeds* (pp. 11–26). Oxford, UK: Blackwell Scientific Publications.

Chang, K.-C. (1970). The beginnings of agriculture in the Far East. *Antiquity*, *44*, 175–185.

Cocannouer, J. A. (1950). *Weeds, guardians of the soil*. New York, NY: Devin-Adair.

Dayton, W. A. (1950). *Glossary of botanical terms commonly used in range research*. USDA Miscellaneous Publication no. 110. (Rev. ed.). Washington, DC: US Government Printing Office.

Dimbleby, G. W. (1967). *Plants and archaeology*. New York, NY: Humanities Press Inc.

Fogg, J. M. (1945). *Weeds of lawn and garden*. Philadelphia, PA: University of Pennsylvania Press.

Fowler, M. L. (1971). The origin of plant cultivation in the Central Mississippi valley: A hypothesis. In S. Struever (Ed.), *Prehistoric agriculture* (pp. 122–128). Garden City, NY: Natural History Press.

Georgia, A. E. (1914). *A manual of weeds*. New York, NY: Macmillan Co.

Ghose, R. L. M., Ghatge, M. B., & Subrahmanyan, V. (1956). *Rice in India*. New Delhi, India: Indian Council Agricultural Research.

Gilmore, M. R. (1930). Dispersal by Indians, a factor in the extension of discontinuous distribution of certain species of native plants. *Michigan Academy of Science, Arts and Letters*, *13*, 89–94.

Halstead, P. (2017). Forest clearance and land use by early farmers in Europe: Insights from north Greek oral history. *Quaternary International*. https://doi.org/10.1016/j.quaint.2017.04.010

Hamilton, A. C. (1972). The interpretation of pollen diagrams from highland Uganda. In E. M. Bakker (Ed.), *Van Zinderen palaeoecology of Africa* (Vol. 7, pp. 45–149). Cape Town, South Africa: A.A. Balkema.

Harlan, H. V. (1929). The weedishness of wild oats. *The Journal of Heredity*, *20*, 515–518.

Harlan, J. R. (1970). *Cynodon* species for grazing and hay. *Herbage Abstracts*, *40*, 233–238.

Harlan, J. R., & deWet, J. M. J. (1965). Some thoughts about weeds. *Economic Botany*, *19*, 16–24.

Harlan, J. R., & deWet, J. M. J. (1969). Sources of variation in *Cynodon dactylon* (L.) Pers. *Crop Science*, *9*, 774–778.

Harlan, J. R., deWet, J. M. J., & Rawal, K. M. (1970). Origin and distribution of the seleucidus race of *Cynodon dactylon* (L.) Pers. var. *dactylon* (Gramineae). *Euphytica*, *19*, 465–469.

Harlan, J. R., & Zohary, D. (1966). Distribution of wild wheats and barley. *Science (Washington, DC)*, *153*, 1074–1080.

Harper, J. L. (1960). *The biology of weeds*. Oxford, UK: Blackwell Scientific Publications.

Hutterer, K. L. (1983). The natural and cultural history of southeast Asian agriculture: Ecological and evolutionary considerations. *Anthropos, 78*, 169–212.

Isely, D. (1960). *Weed identification and control.* Ames, IA: Iowa State University Press.

Johnson, S. (1827). *Dictionary of the English language.* London: Longman, Rees, Orme, Browne, and Green.

King, F. C. (1951). *The weed problem, a new approach.* London: Faber & Faber Ltd.

Klingman, G. C. (1961). *Weed control: As a science.* New York, NY: John Wiley & Sons.

Körnicke, F. (1885). *Die Arten und Varietaten des Getreides. Handbuch des Getreidebaues.* Berlin, Germany: Paul Parey.

Livingstone, D. A. (1984). Interactions of food production and changing vegetation in Africa. In J. D. Clark & S. A. Brandt (Eds.), *From hunters to farmers: The causes and consequences of food production in Africa* (pp. 22–25). Berkeley, CA: University of California Press.

Lundell, C. L. (1938). Plants probably utilized by the old empire Maya of Petén and adjacent lowlands. *Michigan Academy of Science, Arts, and Letters, 24*, 37–56.

Merrill, E. D. (1922). *An enumeration of Philippine flowering plants.* Manila, Philippines: Bureau of Printing.

Mitra, S. K., & Ganguli, P. M. (1932). Some observations on the characters of wild rice hybrids. *Indian Journal of Agricultural Sciences, 2*, 271–279.

Moore, J. H. (1958). *Agriculture in ante-bellum Mississippi.* New York, NY: Bookman Associates.

Morrell, P. L., Williams-Coplin, T. D., Lattu, A. L., Bowers, J. E., Chandler, J. M., & Patterson, A. H. (2005). Crop-to-weed introgression has impacted allelic composition of johnsongrass populations with and without recent exposure to cultivated sorghum. *Molecular Ecology, 14*, 2143–2154. https://doi.org/10.1111/j.1365-294X.2005.02579.x

Moseley, E. L. (1930). Some plants that were probably brought to northern Ohio from the west by Indians. *Michigan Academy of Science, Arts, and Letters, 13*, 169–172.

Muenscher, W. C. (1946). *Weeds.* New York, NY: Macmillan.

Murray, J. A. H., Bradley, H., Cragie, W. A., & Onions, C. T. (1961). *Oxford English dictionary* (Vol. *12* + *supplements*). Oxford, UK: The Clarendon Press.

Norris, K., Asase, A., Collen, B., Gockowski, J., Mason, J., Phalan, B., & Wade, A. (2010). Biodiversity in a forest-agricultural mosaic—the changing face of West African rainforests. *Biological Conservation, 143*, 2341–2350.

Ohadi, S., Hodnett, G., Rooney, W., & Bagavathiannan, M. (2017). Gene flow and its consequences in *Sorghum* spp. *Critical Reviews in Plant Sciences, 36*(5–6), 367–385. https://doi.org/10.1080/07352689.2018.1446813

Portères, R. (1957). Paysages floristiques des parcours cultureaux en Afrique topicale. *Compte rendu des séances du Société biogéographique, 294*, 16–20.

Pritchard, T. (1960). Race formation in weedy species with special reference to *Euphorbia cyparissias* L. and *Hypericum perforatum* L. In J. L. Harper (Ed.), *The biology of weeds* (pp. 61–66). Oxford, UK: Blackwell Scientific Publications.

Rademacher, B. (1948). Gedanken über Begriff und Wesen des "Unkrautes.". *Pflanzenkrankheiten und Pflanzenschutz, 55*, 3–10.

Ramiah, K., & Ghose, R. L. M. (1951). Origin and distribution of cultivated plants of South-Asia: Rice. *Indian Journal of Genetics and Plant Breeding, 11*, 7–11.

Robbins, W. W., Crafts, A. S., & Raynor, R. N. (1942). *Weed control*. New York, NY: McGraw Hill.

Salisbury, E. (1961). *Weeds and aliens*. London: Collins.

Spalding, T. (1844). Letter to editor. *American Agriculturist, 3*, 335.

Watson, P. J. (1989). Early plant cultivation in the eastern woodlands of North America. In D. R. Harris & G. C. Hillman (Eds.), *Foraging and farming: The evolution of plant exploitation* (pp. 555–571). London: Unwin Hyman.

Wilkes, H. G. (1967). *Teosinte: The closest relative of maize*. Cambridge, MA: Bussey Institute, Harvard University.

Wodehouse, R. P. (1960). Weed. *Encyclopaedia Britanica, 23*, 477–479.

Yabunó, T. (1961). *Oryza sativa* and *Echinochloa crus-galli* var. *oryzicola* Ohwi. *Seiken Ziho, 12*, 29–34.

Yarnell, R. A. (1965). Implications of distinctive flora of Pueblo ruins. *American Anthropologist, 67*, 662–674.

Zohary, D. (1971). Origin of south-west Asiatic cereals: Wheats, barley, oats, and rye. In P. H. Davis (Ed.), *Plant life of southwest Asia* (pp. 235–263). Edinburgh, Scotland: Botanical Society of Edinburgh.

5

Classification of Cultivated Plants

THEOPHRASTUS (CA 2370–CA 2287 BP)

So out of the ground the Lord God formed every beast of the field and every bird of the air, and brought them to the man to see what he would call them; and whatever the man called every living creature, that was its name.

Gen. 2:19 Revised Standard Version

Source: Chapter title page sketches by Patricia J. Scullion, ASA/CSSA Headquarters Office.

Harlan's Crops and Man: People, Plants and Their Domestication, Third Edition.
H. Thomas Stalker, Marilyn L. Warburton, and Jack R. Harlan.
© 2021 American Society of Agronomy, Inc. and Crop Science Society of America, Inc.
Published 2021 by John Wiley & Sons, Inc.
doi:10.2135/harlancrops

LINNEAUS (1707–1778)
Botanists have generally neglected cultivated varieties as beneath their notice.
Darwin (1897)

Botanical Descriptions and Names

The primary purpose of classification is to reduce the number of items to manageable proportions. If there were only 100 plants on earth, we could assign a name, number, or other symbol to each and deal with them individually. Because there are millions on millions of plants of thousands on thousands of different kinds, for convenience we group them so that we can deal with a reduced number of categories of plants. Obviously, it would do little good to group plants at random; if the reductional system is going to work, we must group like with like.

The hardware merchant puts his bolts into separate bins. For the arrangement to be convenient, he must classify them. First, he sorts out the major classes; stove bolts, machine bolts, carriage bolts, and so on. Each class is then divided by diameter and within each of these categories divided again by length. Ultimately, each bin contains items that are essentially identical. Plants are not so easy to sort, but the purpose and the method are about the same. The bins will not contain identical plants, but we would like them to contain individual plants that resemble each other more than they resemble individuals in other bins.

To classify like with like, we must have methods of description. We must decide which characters are useful in grouping plants and which characters are too inconsistent to be helpful. We must not only describe the individual to be classified, but the category to which it will be referred. Botanical description is basic to plant classification, but there is as much art as science to it. Taxonomists differ enormously in their ability to find suitable characters for groups and in their ability to describe simply, clearly, and unambiguously. The introduction of numerical taxonomy and computers has not improved the situation noticeably, although DNA sequencing is starting to.

As we describe categories of plants, we inevitably find that some groups resemble each other more than they resemble other groups. Broad genetic affinities are established that reveal evolutionary history in a rather general way. If our taxonomy is a good one, plants assigned to one genus are more nearly related to each other than to plants assigned to another genus, and similarly for plants belonging to a family. No matter how the classification is arranged, however, we encounter intermediates that do not fit well into one group or another and we find anomalous groups that do not seem to belong to any group in particular.

Finally, classification provides an opportunity to give something a name. Names are very important, and most societies have naming ceremonies that are taken very seriously. In some societies, in fact, the real name of a person is kept secret and a different every day name used instead. The reason is the belief that knowledge of the true name confers a terrible destructive power. One can kill an enemy by simply pronouncing his real name. A given name confers a very special identity on a person or a thing, and verifies its existence.

We found this to be true of plants as well. In studies of the biosystematics of *Cynodon,* we found out very early that we had in our collection taxa that had not been named. We accumulated a great deal of information about geographic distribution, morphological variation, genetic affinity, chromosome pairing in hybrids, crossability, fertility, sterility, and so on, but we could not publish any of it until the taxa were officially described. Botanically, these plants did not exist until names were given.

Classifications, then, lump individuals into groups so that we can deal with categories of plants instead of vast numbers of individuals. They reveal genetic affinity and evolutionary history, and they describe and give names to plants. Taxonomy is, pragmatically, a science of convenience. It makes it possible for man to deal rationally with the vast arrays of variation found in the natural world.

The most important categories for plant classification are family, genus, and species, while infraspecific categories are especially useful in cultivated plants. Families may be grouped into order, orders into phyla, and so on, but for our purpose, the lower groupings are the most useful. Families are usually defined by floral morphology together with a few basic vegetative traits. Using our 30 major crops as a sample (Table 2.1 in Chapter 2), we find wheat, maize, rice, barley, sugarcane, sorghum, oats, rye, and the millets in the grass family, Gramineae; soybean, bean, peanut, and pea are in the legume family, Leguminoseae; and potato and tomato are in the Solanaceae; and rape and cabbage in the Cruciferae. Each of the others belongs to a different family. While the grass family dominates the list and the food supply, it is one of the most difficult to treat taxonomically because most of the flower parts have been reduced to scales or rudiments or are not present. A student of agriculture should have some familiarity with the family, however, not only because it furnishes the cereals, but because grasslands cover much of the surface of the earth. Grasslands protect land from erosion, provide grazing and fodder for animals, sods, turf, and lawns for landscape and recreation. In like manner, the legume family not only provides the pulses, which are important nutritionally, but many are valuable trees, and pasture and forage legumes take part in fixing nitrogen to enrich the soil and enhance production.

Problems of Formal Taxonomy

As Darwin (1859) pointed out (see p. 102), botanists have had little to do with classification of cultivated plants. This may not be so much because such plants are beneath their notice as because the traditional taxonomist is bewildered and confused by cultivated plants and doesn't know what to do with them. When botanists do try to classify cultivated plants, remarkably erratic results can be expected.

The inconsistencies and lack of agreement among taxonomists dealing with the same materials are remarkable, to say the least, and are even more striking when the treatments of different crops are compared. Past confusion and disagreement extended over the generic, specific, and infraspecific levels. For example, in the wheats, Percival (1921) listed two species, Bowden (1959) three, and Jakubziner and Dorofeev (1968) 24, but all were classifying essentially the same materials. Snowden (1935, pp. 221–255) used 31 species of cultivated *Sorghum* alone, in addition to the wild and weedy ones that are fully compatible genetically with the domesticated types; Jakushevsky (1969) reduced these to nine, and deWet and Huckabay (1967) to one. Bukasov (1933) had well over 200 species in the Tuberarium section of *Solanum*; Hawkes (1963, pp. 76–181) reduced these to about half that many, yet retained 64 species in section Tuberosa, in which the taxa can be intercrossed and in which there is very little genomic differentiation despite a fairly extensive polyploid series.

New DNA sequence evidence is now clearing up past confusions. For example, some taxonomists assigned teosinte to the genus *Euchlaena*, and some to *Zea*, but now, all teosinte species belong to the genus *Zea*, and some [*Zea mays* subsp. *huehuetenangensis* (H. H. Iltis & Doebley) Doebley, *Z. mays* subsp. *mexicana* (Schrad.) H. H. Iltis, and *Z. mays* subsp. *parviglumis* H. H. Iltis & Doebley], belong to the same subspecies as domesticated maize, *Z. mays* subsp. *mays*. *Aegilops*, which had been assigned to the genus *Triticum* by some taxonomists, is now known to be an independent genus, although one of the three ancestral contributors to hexaploid breadwheat.

The number of examples of this kind can be multiplied many times. Faced with this sort of vacillation and indecision among taxonomists, the people who deal with cultivated plants the most (geneticists, agronomists, horticulturalists, and foresters) have developed their own informal and intuitive classifications, based on experience, as to what constitutes useful groupings. They will continue to use their own systems no matter what the taxonomist does or does not do. There is more involved here, however, than the usual differences in judgment between "splitters" and "lumpers." First, cultivated plants *are* different from wild ones and require special taxonomic treatment; and second, there have been no guidelines for consistent group-ings of related taxa according to the degree of relationship.

How do cultivated plants differ from wild species? Their variation pat-terns are different. Darwin (1859) opened his book on the origin of spe-cies with a discussion of these differences in both plants and animals. Broccoli, Brussels sprouts, cabbage, cauliflower, kale, and kohlrabi do not look much alike, but belong to the same species in a biological sense. They can be crossed readily and their hybrids are fertile. In appearance they are as different as collies, terriers, Great Danes, and chows, and these all belong to a single species as well. Darwin considered the enor-mous arrays of variation found among domestic breeds of cattle, horses, sheep, pigeons, and chickens, and among wheat, roses, peas, dahlias, iris, carrots, and so on. The morphological differences among genetically related breeds are simply of a different order of magnitude from those which are found in wild species.

The science of genetics was not developed in Darwin's time and he thought that variation was induced by changes in environment, but it is true that conditions of domestication lead to a wider range of variation. Strange and bizarre forms that might appear in nature are usually promptly pruned out by natural selection, but in cultivated plants the strange and bizarre are likely to be precisely the ones to be selected and propagated. Most of our cultivars are biological "monsters" that could not survive in the wild but are cultivated by man because they please him in some way.

Furthermore, man has been very active in manipulating gene pools through repeated introduction or migration followed by natural or artificial hybridization. The germplasm of domesticated plants has been repeatedly and periodically stirred (Harlan, 1966, 1969, 1970). The environment provided by man has been artificial, unstable, and often extensive geographically. Selection pressures have been very strong but biologically capricious and often in diverse directions. The result is an enormous amount of conspicuous phenotypic variation among genetically very closely related forms.

Faced with this situation, the traditional taxonomist tends to overclassify. He finds conspicuous either/or characters often without intermediates and frequently bases species on them. These characters may be controlled by one or a few genes and have little biological significance. Too many species and too many genera are named and then, to accommodate the enormous variability remaining, unreasonable numbers of infraspecific classes may be established.

de Candolle (1867, 1883) recognized the situation rather clearly and objected strongly to the application of Latin names to "horticultural productions." It is a question of fundamental taxonomy: What are the most useful characters for the separation of groups? In cultivated plants, it is the plant breeders who will make the most use of classification of their germplasm. The most useful characters will be morphological ones that can be recognized at a glance, since it is often necessary to deal with large numbers of plants. Infraspecific classifications should be simple and repeatable so that plant breeders in different parts of the world can use the same system and understand each other. With the advent of cheap and rapid sequencing, the genomic sequence becomes a reliable character for grouping, which avoids the problem of relying on extreme phenotypic diversity regulated by a few genes.

The Gene Pool System

Species

There is too little agreement among taxonomists as to species limits. The most important limits for the plant breeder are those that deal with genetic compatibility. To provide a genetic perspective and genetic focus for cultivated plants, Harlan and deWet (1971) proposed three informal categories: (a) primary gene pool, (b) secondary gene pool, and (c) tertiary gene pool (Figure 5.1).

Primary Gene Pool (GP-1)
This corresponds to the traditional concept of the biological species. Among forms in this gene pool, crossing is easy, hybrids are generally fertile with good chromosome pairing, gene segregation is approximately normal, and gene transfer is generally simple. The biological species almost always includes spontaneous races (wild and/or weedy) as well as cultivated races. To make this clear they proposed that the species be divided into two subspecies: (a) subspecies A, which includes the cultivated races, and (b) subspecies B, which includes the spontaneous races.

Secondary Gene Pool (GP-2)
This includes all biological species that will cross with the crop and approximates an experimentally defined coenospecies. Gene transfer is possible, but one must struggle with those barriers that separate biological species. Hybrids tend to be sterile, chromosomes pair poorly or not at all, some hybrids may be weak and difficult to bring to maturity, and recovery of desired types in advanced generations may be difficult. The gene pool is available to be utilized, however, if the plant breeder or geneticist is willing to make the required effort.

Tertiary Gene Pool (GP-3)
At this level, crosses can be made with the crop, but the hybrids tend to be anomalous, lethal, or completely sterile. Gene transfer is either not possible with known techniques or else rather extreme or radical measures are required (e.g., embryo culture, grafting or tissue culture to obtain hybrids, doubling chromosome number, or using bridging species to obtain some fertility). The value of GP-3 is primarily informational; it defines the extreme outer limit of potential genetic reach. If a cross can be made at all, however, there is always a chance that some technique will be discovered that will make it possible to use germplasm in the tertiary gene pool. Since very few people have worked at this level, GP-3 is likely to be rather ill defined, but will be better known as information accumulates.

Perhaps the most powerful traditional tool now known for introducing genes from GP-3 in a crop is the use of complex hybrids. For example, Russian scientists under Tsitsin (1962) tried for many years to cross *Elymus* with *Triticum*. After a great many failures they finally obtained a few sterile hybrids by embryo culture techniques earlier developed by Ivanovskaya (1946). Later they found that the use of an *Agropyron* × *Triticum* derivative as a female parent permitted a straightforward incorporation of *Elymus* germplasm without using special techniques employed by Soulier (1945), and similar results were obtained using amphiploid wheat × rye derivatives first developed by Pissarev and Vinogradova (1945). Hybrids between *H. bulbosum* L. and *H. vulgare* L. at both the diploid and tetraploid levels are generally completely sterile, but Schooler (1967, 1968) was able to incorporate germplasm of *H. bulbosum* into *H. vulgare* and recover female fertility by way of a complex cross involving *Hordeum jubatum* L. and *Hordeum compressum* Griseb. as well. deWet et al. (1970), working with *Zea* and *Tripsacum*, were able to incorporate teosinte and *Tripsacum* for the first time only by way of maize×*Tripsacum* hybrids. Previously, Harlan and deWet (1963) had similar experiences with wide crosses in *Bothriochloa* and some of the widest crosses in sugarcane have involved complex hybrid materials (Price, 1957).

The secondary gene pool might outline groups that would be acceptable to some taxonomists as generic limits, but the tertiary gene pool may extend too far. Price (1957), after reviewing the wide crosses with sugarcane, remarked:

> The results from hybridizing *Saccharum* species and their allies can be regarded only as fragmentary. Yet they suggest that eventually it may be necessary to return certain species of *Erianthus, Narenga* and *Sclerostachya* to *Saccharum*. On the other hand, despite the undoubted validity of sugarcane × *Sorghum* crosses and the possibility that hybrids from sugarcane × maize may yet be found genuine, one can scarcely imagine a genus which would include species presently assigned to *Saccharum, Sorghum,* and *Zea.* Nor is a taxonomist likely to accept a genus including *Triticum, Aegilops, Secale, Haynaldia,* and only some species of *Agrypyron* and of *Elymus.*

GP-3 describes the extreme outer limit of the potential gene pool of a crop. It is not a taxonomic unit in the conventional sense and the terms primary, secondary, and tertiary gene pools are not proposed as formal taxonomic categories. They are simply guides for placing classification into genetic perspective. The system is shown diagrammatically in Figure 5.1, and examples of the peanut group are shown in Figure 5.2 and rice in Figure 5.3.

The system is flexible and subject to change with more effort and more research. In the first edition of this book, I claimed that soybean had neither a GP-2 nor GP-3. Since then hybrids have been made with wild perennial relatives. Since embryo rescue was used, the relationship seems to be at the GP-3 level. Barley also seems closer to wheat than we once thought, and maize has by now been crossed with all or nearly all species of *Tripsacum* and gene exchange is possible. A biological species, GP-1, may have fuzzy boundaries because some hybrids may show ranges of fertility. Genetic compatibility also can be variable. Our purpose in proposing the gene pool

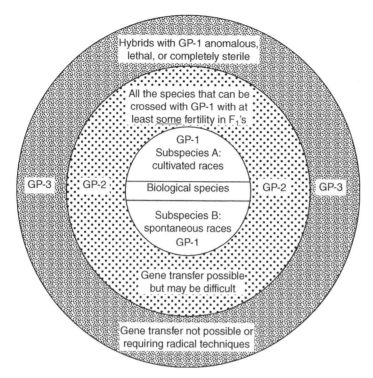

Figure 5.1   Schematic diagram of primary gene pool (GP-1), secondary gene pool (GP-2), and tertiary gene pool (GP-3). *Source:* Harlan and deWet (1971). Reproduced with permission from John Wiley & Sons.

system was to provide a rational basis for comparative taxonomies. The system has received fairly wide acceptance.

Genetic engineering is opening up new possibilities that permit genetic transfer from beyond the GP-3. To account for the expanded potential of using germplasm outside the genus, some researchers have added a fourth gene pool (GP-4) to include all organisms for which gene manipulation technologies can potentially be inserted into the respective cultivated genome. Not all plants are easy to work with this kind of manipulation. However, as techniques advance, we have found ways to insert genes from unrelated plants, animals, fungi, bacteria, and viruses. If a gene can be isolated as a DNA fragment and a suitable vector is available, transfers need not depend on species relationships. Contributions to agriculture to date have mainly been genes of greatest economic benefit, but genes improving stress tolerance and nutritional content, and others of benefit to the consumer or the environment, are becoming more common.

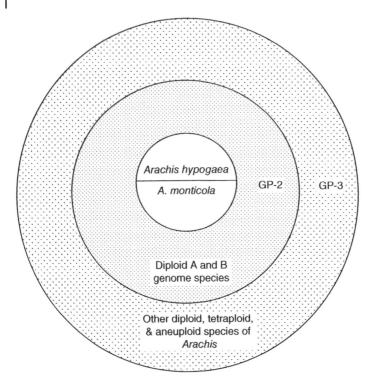

Figure 5.2 Gene pools of *Arachis*. The cultivated species is an allotetraploid species with A & B genomes. The secondary gene pool contains nearly 30 diploid species with either an A or a B genome, and the tertiary gene pool has more than 50 additional species of which most are diploid but polyploid wild species also exist. There are many distinct genomes, generally following sectional classifications, in the tertiary gene pool and species are reproductively isolated from *A. hypogaea*.

The problem of polyploidy must, of course, be dealt with. There may be no solution that would apply to all crops, but as a general guideline it was proposed that separate gene pools be recognized for different ploidy levels. This would not apply to artificial or induced polyploids; for example, tetraploid barley and tetraploid maize would be included in the diploid gene pools for barley and maize. On the other hand, *Sorghum halepense* ($2n = 40$) would be separated from *S. bicolor* (L.) Moench ($2n = 20$). In wheat, four cultivated gene pools are recognized representing three ploidy levels. These are separated from the *Triticum timopheevii* (Zhuk.) Zhuk.—*T. timopheevii* subsp. *armeniacum* (Jakubz.) Slageren. group on the basis of chromosome pairing and sterility barriers. The genetic barriers due to polyploidy are not always strong and gene transfer across ploidy levels may be rather easy. The barriers are there, however, and it is generally useful to indicate their presence by

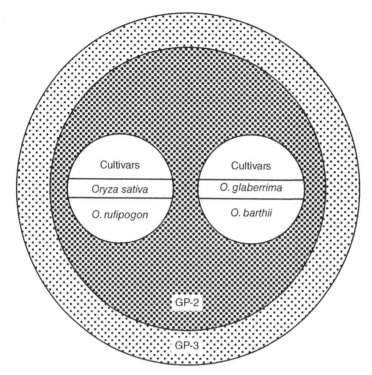

Figure 5.3 Gene pools of rice with two rice domesticates with separate primary gene pools. The secondary gene pool contains 6 diploid species with the same AA genome, but the entire genus contains 14 more diploid and tetraploid species with genome types BB, CC, BBCC, CCDD, EE, FF, GG, and HHJJ, which comprise the tertiary gene pool.

providing separate epithets. For certain crops, like potato and sugarcane, this may not be appropriate and each crop must be treated as a separate case.

Subspecies

Formal taxonomy has failed most conspicuously at the infraspecific level in cultivated plants. Many systems have been proposed and there is little agreement among specialists. There is a strong tendency to overclassify and give formal categories to groups of cultivars that have little or no genetic integrity. In the proposal of Harlan and deWet (1971), the biological species (GP-1) would first be partitioned into two subspecies, one containing the cultivars, the other including the spontaneous forms. At this point it is recommended that all formal taxonomy be abandoned to permit the use of the informal systems used by those who work with cultivated plants professionally. The term "varieties" would be avoided as a botanical term because it is too easily confused with agronomic and horticultural varieties (cultivars) and is not especially appropriate for cultivated plants in any case. The infraspecific categories proposed are:

Species
1. Subspecies A. The cultivated races
   2. Race
      3. Subrace
         4. Cultivar
            5. Line, clone genotype

1. Subspecies B. The spontaneous races
   2. Race
      3. Subrace

A classification of cultivated plants does not require more divisions than this. A reasonable amount of variation must be allowed for both race and subrace, but more infraspecific categories tend to destroy the purpose of classification.

Race
In the classification proposed, it is necessary to have some understanding of what constitutes a race. A race not only has a recognizable morphological identity but also is a biological unit with some genetic integrity. A race originated in some geographic region at some time in the history of the crop. As a biological unit it is not as clearly separable as species but has a distinct cohesion of morphology, geographical distribution, ecological adaptation, and frequently breeding behavior.

It is understood, of course, that racial differentiation is not always clear-cut. There are ill-defined races, hybrid races, races in the process of formation, and complex races made up of derivatives of two or more races. This is the very reason that formal categories have failed and why informal systems based on the experience of plant breeders are to be preferred. Ultimately, a race becomes simply a useful group of cultivars, but the group is most useful when it has a biological basis.

One may also express the presumptive derivations of groups by very simple combinations of the basic elements of variation. As an example, Harlan and deWet (1972) classified the cultivated sorghums into five basic races: bicolor, guinea, caudatum, kafir, and durra. There are, however, clearly identifiable, intermediate races involving all combinations of these basic races: kafir-caudatum, most modern American grain sorghums; durra-caudatum, the kauras of Nigeria and similar subraces; and guinea-kafir, the common shallu sorghum of India; guinea-caudatum, common in Nigeria, Chad, and

Table 5.1   Classification of sorghum.

Species:	*Sorghum bicolor*
Subspecies:	*S. bicolor* subsp. *bicolor*
Race:	Guinea
Subrace:	guineense
Cultivar:	Sabba Bibi
Line:	A selection from the cultivar
Genotype:	An individual plant or homozygous line selected from the cultivar
Clone:	An individual plant selected from the cultivar and propagated asexually by cuttings, tissue culture, apomixis, and so on.

Sudan. The hyphenated names imply exactly what the race appears to be on a morphological and distributional basis and thus provide useful information.

Subrace

The subrace is simply a convenient division of a race. It must be reasonably recognizable to be convenient and it may or may not be appropriate to divide a race into subraces. In most cultivated plants there is such a continuous range of variation from subrace to subrace and sometimes from race to race that there is no useful purpose to be served by very fine divisions in the classification.

It must be emphasized that races and subraces are not intended to be formal categories and they are not to be italicized. Racial classifications should be flexible and subject to change as more experience and genetic data becomes available, and it is recommended that rigid rules for their applications be avoided. Indeed the entire system of gene pools as proposed by Harlan and deWet (1971) is without formal terminology in the usual sense. Partitions of the gene pool are designed to give consistency to classifications already available and suggestions for infraspecific categories are designed to permit use of the informal systems that have been found by experience to be useful.

Evolutionary Implications

The system permits us to compare different crops, on a more or less genetically equivalent basis. When this is done, it becomes apparent that most crop evolution takes place within the primary gene pool. Domestication very seldom leads to speciation despite the many classifications that provide separate epithets for domesticates. Genetically, domesticated races belong to the same biological species as their wild progenitors and are fully compatible with them when hybridized. There are very few crops for which wild races have not yet been identified and most of these have been little studied.

Among the cereals, the only exception is hexaploid bread wheat, which is not known in the wild. The gene pool apparently arose after domestication of tetraploid wheat through the addition of the D genome of *Aegilops squarrosa*. It is one of the very few demonstrable cases of a new gene pool being generated under domestication. Various degrees of genomic modification have taken place in other polyploids, such as potato, sweet potato, and sugarcane. Nevertheless, with very few exceptions, crop evolution has operated at the infraspecific level.

We know that polyploidy is a form of quantum evolution in that it takes only one or two generations to pass from one ploidy level to another. In the absence of polyploidy, evolution is much slower, and it appears that the processes of domestication have not been operating long enough to bring about speciation. According to our archaeological information we are dealing with between 10,000 and 12,000 generations for our older annual crops and many fewer for perennials. Either this is not enough to establish substantial genetic barriers or the sporadic crossing between wild and cultivated races has prevented establishment of separate gene pools. The morphological modifications in this period of time, however, have sometimes been spectacular.

In many cases, we do not know much about the role of secondary gene pools in crop evolution. There are reasons to suspect that *Aegilops, Secale, Agropyron,* and *Haynaldia* have contributed something to wheat; that *Oryza longistaminata* has contributed something to African rice and even to Asian rice in Africa (Chu & Oka, 1970); and that *Saccharum spontaneum* L. has introgressed into sugarcane. It has repeatedly been suggested that *Tripsacum* has contributed to maize evolution, but there has been no direct evidence of gene introgression. Interactions between species in the primary and secondary gene pools can be produced artificially, but it is much more difficult to establish where genetic contributions to the crop's evolution occurred under natural conditions.

Whatever infusions of germplasm may have taken place from the secondary gene pool into the primary gene pool, they are of such a nature that new gene pools are seldom formed. Wild and cultivated emmers are fully compatible, as are wild and cultivated rices and teosinte and cultivated maize. From experience with several hundred crop species, it seems that primary gene pools have not been disrupted by whatever contributions might have come from secondary gene pools.

Conclusions

Classification of cultivated plants is necessary to deal with the vast arrays of diversity and to understand patterns of variation within each crop. Conventional taxonomy as used for wild species tends to overclassify and provide too many categories. An informal system based on gene pools, races, and subraces is proposed. Plant breeders and others who work with cultivated plants professionally will continue to use their own intuitive classifications because they are practical and work. It is recommended that Latin names not be used for categories below the subspecies level.

References

Bowden, W. M. (1959). The taxonomy and nomenclature of the wheats, barleys and ryes and their wild relatives. *Canadian Journal of Botany, 37*, 657–684. https://doi.org/10.1139/b59-053

Bukasov, S. M. (1933). The potatoes of South America and their breeding possibilities. *Trudy po Prikladnoi Botanike, Genetike i Selektsii, Suppl. 58*, 192.

Chu, Y.-E., & Oka, H.-I. (1970). Introgression across isolating barriers in wild and cultivated *Oryza* species. *Evolution, 24*, 344–355. https://doi.org/10.1111/j.1558-5646.1970.tb01766.x

Darwin, C. R. (1859). *On the origin of species by means of natural selection.* London: J. Murray.

Darwin, C. R. (1897). *The variation of animals and plants under domestication* (2nd ed.). London: J. Murray. https://doi.org/10.5962/bhl.title.127981

de Candolle, A. (1867). *Lois de la nomenclature botanique.* Paris, France: V. Masson + Fils.

de Candolle, A. (1883). *Nouvelles remarques sur la nomenclature botanique.* Geneva, Switzerland: H. Georg.

deWet, J. M. J., & Huckabay, J. P. (1967). The origin of *Sorghum bicolor.* II. Distribution and domestication. *Evolution, 21*, 787–802.

deWet, J. M. J., Lambert, R. J., Harlan, J. R., & Naik, S. M. (1970). Stable triploid hybrids among *Zea-Tripsacum-Zea* backcross populations. *Caryologia, 23*, 183–187.

Harlan, J. R. (1966). Plant introduction and biosystematics. In K. J. Frey (Ed.), *Plant breeding* (pp. 55–83). Ames, IA: Iowa State University Press.

Harlan, J. R. (1969). Evolutionary dynamics of plant domestication. *Japanese Journal of Genetics, 44*, 337–343.

Harlan, J. R. (1970). The evolution of cultivated plants. In O. H. Frankel & E. Bennett (Eds.), *Genetic resources in plants—their exploration and conservation* (pp. 19–32). Oxford, England: Blackwell Scientific Publications.

Harlan, J. R., & deWet, J. M. J. (1963). The compilospecies concept. *Evolution, 17*, 497–501. https://doi.org/10.1111/j.1558-5646.1963.tb03307.x

Harlan, J. R., & deWet, J. M. J. (1971). Toward a rational classification of cultivated plants. *Taxon, 20*, 509–517.

Harlan, J. R., & deWet, J. M. J. (1972). A simplified classification of cultivated sorghum. *Crop Science, 12*, 172–176. https://doi.org/10.2135/cropsci197 2.0011183X001200020005x

Hawkes, J. G. (1963). *A revision of the tuber-bearing Solanums* (2nd ed.). Edinburgh, Scotland: Scottish Plant Breeding Station.

Ivanovskaya, E. V. (1946). Hybrid embryos of cereals grown on artificial nutrient medium. *Izvestiya Akademii Nauk SSSR, Seriya Biologicheskaya, 54*, 445–448.

Jakubziner, M. M., & Dorofeev, S. F. (1968). World wheat resources in service of Soviet breeding. *Trudy po Prikladnoi Botanike, Genetike i Selektsii, 39*, 65–79.

Jakushevsky, E. S. (1969). Varietal composition of sorghum and its use for breeding. *Trudy po Prikladnoi Botanike, Genetike i Selektsii, 41*, 148–178.

Percival, J. (1921). *The wheat plant, a monograph*. London: Duckworth and Co.

Pissarev, V. E., & Vinogradova, N. M. (1945). Trigeneric hybrids elymus×wheat×rye. *C. R. (Doklady) Akademii nauk SSSR: Comptes rendus de l'Academie des Sciences de l'URSS, 49*, 218–219.

Price, S. (1957). Cytological studies in *Saccharum* and allied genera. III. Chromosome numbers in interspecfic hybrids. *Botanical Gazette, 118*, 146–159. https://doi.org/10.1086/335939

Schooler, A. B. (1967). A form of male sterility observed in *Hordeum* L. hybrids. *Agronomy Abstracts, 17*. Madison, WI: ASA

Schooler, A. B. (1968). Cytoplasmic sterility and the production of hybrid barley seed. *North Dakota Farm Research, 25*, 10–11.

Snowden, J. D. (1935). A classification of the cultivated sorghums. *Bulletin of Miscellaneous Information (Royal Botanic Gardens, Kew), 1935*(5), 221–255.

Soulier, E. J. (1945). A composite perennial *Elymus*-Wheat-*Agropyrum* hybrid. *Izvestiya Akademii Nauk SSSR, Seriya Biologicheskaya, 47*, 578–579.

Tsitsin, N. V. (1962). The significance of wide hybridization in the evolution and production of new species and forms of plants. In N. V. Tsitsin (Ed.), *Wide hybridization of plants* (Vol. *1*, pp. 2–30). Jerusalem, Israel: Israel Program for Scientific Translations.

6

The Dynamics of Domestication

The unlike is joined together, and from differences results the most beautiful harmony, and all things take place by strife.
Heraclitus, fifth century BC (Patrick, 1889)

Domestication of Seed Crops

Cereals

Greek philosophers were fascinated by the problem of change. How could a thing become something other than what it is—unless it was secretly the

Source: Chapter title page sketches by Patricia J. Scullion, ASA/CSSA Headquarters Office.

Harlan's Crops and Man: People, Plants and Their Domestication, Third Edition.
H. Thomas Stalker, Marilyn L. Warburton, and Jack R. Harlan.
© 2021 American Society of Agronomy, Inc. and Crop Science Society of America, Inc.
Published 2021 by John Wiley & Sons, Inc.
doi:10.2135/harlancrops

other thing all the time? Heraclitus, a philosopher from Ephesus, thought that everything in the universe was in a constant state of flux and that the only way stability could be achieved would be through a balance of opposing forces. This is fundamentally an evolutionary dynamic. In a variable genetic field, populations may become stable only through centripetal selection pressures. If pressures change one way or another, the populations will change accordingly.

Domestication is an evolutionary process operating under the influence of human activities. Since it is evolutionary, we would expect a relatively slow and incipient domestication to forms that differ more and more from the progenitors. We would also expect that it would be difficult to separate initial steps in domestication from truly wild or weedy forms, and that it would be impossible to predict how far the process might extend into the future. As a matter of fact, we are domesticating plants all the time and find that the processes can be simple and rapid, depending on the species selected.

We visualize the domestication of cereals as beginning with the harvest of wild grass seeds. We have seen that this was common and widespread among gathering peoples and continues to this day. As long as human activity is confined to harvesting, any genetic effect on wild populations is likely to be negligible. It is the seeds that escape the harvester that produce the next generation, and if there is any selection pressure at all it would be in favor of such wild-type characters as shattering, indeterminate growth with maturation over a long period of time, seed dormancy, and so on. As soon as man starts to plant what he has harvested, the situation changes drastically.

Now there are two populations, one spontaneous, the other being harvested and planted, and the selection pressures are in opposite directions. The seeds that are harvested are those that contribute to the sown population, and any modification that would enhance seed recovery and competition in the new environment would be selected favorably. Automatic selection for interrelated syndromes of characteristics is set up immediately (Table 6.1).

Selection Associated with Harvesting

Of all the adaptations that separate wild from cultivated cereals, the nonshattering trait of cultivated races is the most conspicuous. It is taxonomically the most diagnostic in separating domestic subspecies from spontaneous subspecies and is crucial in establishing the disruptive selection that effectively maintains separation of the two kinds of populations. Most of the seeds that shatter escape the harvest.

In most of the major cereals, it seems that shattering is under relatively simple one- or two-gene control. Appearance may be somewhat deceiving.

Table 6.1 Adaptation syndromes resulting from automatic selection due to planting harvested seed of cereals.

I. Selection pressures associated with harvesting

  A. Increase in percent seed recovered

  Adaptations: 1. Nonshattering

  2. More determinant growth

  a. Growth habit I: Cereals whose wild races have lateral seed-bearing branches, for example, maize, coix, sorghum, pearl millet. There is a trend toward apical dominance resulting in fewer inflorescences, larger inflorescences, larger seed, greater daylength sensitivity, and more uniform ripening

  b. Growth habit II: Cereals with unbranched culms, for example, barley, emmer, rye, einkorn, rice. There is a trend toward more synchronous tillering and uniform whole plant maturation

  B. Increase in seed production

  Adaptations: 1. Increase in percent seed set

  2. Reduced or sterile flowers become fertile

  3. Increase in inflorescence size, especially in maize, sorghum, pearl millet

  4. Increase in number of inflorescences especially wheat, barley, rice, and so on

II. Selection pressures associated with seedling competition

  A. Increase in seedling vigor

  Adaptations: 1. Greater seed size

  2. Lower protein, higher carbohydrate

  B. More rapid germination

  Adaptations: 1. Loss or reduction of germination inhibitors

  2. Reduction in glumes and other appendages

III. Selection pressures associated with tillage and other disturbances result in the production of weed races

  Adaptations: 1. Plants competitive with cultivated races, but

  2. Retain the shatter habit of wild races

Source: Adapted from Harlan et al. (1973).

Intermediate semishattering forms are known in all cases but are relatively uncommon. Such a condition is not well adapted to either cultivation or spontaneous conditions. Strong disruptive selection for either one state or the other will produce at least the appearance of simple either-or inheritance. The establishment of nonshattering traits is genetically one of the easiest and simplest steps in the entire process of domestication.

There are cases, however, where semishattering may have some advantage. We have already mentioned the semishattering Ethiopian weed oats and weed rye, where some of the seed falls and infests the ground and the rest is harvested and planted with the crop it infests. The nonshattering species of Ethiopian oats, *Camelina sativa* and *Bromus secalinus*, also have been mentioned as weeds that are adapted to being harvested and planted along with their companion crops even though the farmer may not want them. Such examples seem to indicate that nonshattering is a trait that automatically shows up in seed crops without intentional selection.

Wilke et al. (1972) suggested that the tough rachis character might not show up if the Native North American's method of harvesting was used. Native Americans harvested most herbaceous seeds by beating them into a basket with a paddle. This is a radically different treatment from that of cutting seeds heads with a sickle or stone knife as was done in the Near East, North Africa, and Australia. If the seeds are knocked into a basket, the fragile spike would be an advantage instead of a disadvantage. This may explain why few cereals were domesticated in the Americas even though many grasses were harvested and some were planted (Chapter 1). Hillman and Davies (1990) studied domestication of wild wheats and barley experimentally and were convinced that the method and time of harvest for selection of shatter resistance is very important.

The well-known "sunflower effect" applies to cereals with lateral seed-bearing branches, such as maize, sorghum, and pearl millet. Wild and weedy sunflowers have many branches bearing a large number of small heads. The ultimate of domesticated types are the mammoth "Russian" cultivars, which have single unbranched stalks bearing enormous single, terminal heads. Maize has followed the same path. The maize progenitor (teosinte) has a branching system in the axils of several leaves on each stem and each branching system includes several small two-ranked fragile ears, each enclosed in a husk. Early maize, which is well represented in archaeological sites from the southwestern United States to southern Mexico, apparently had clusters of very small, four-ranked (i.e., eight-rowed), mostly nonfragile ears in the axils of several leaves of each stem. From this condition there was a gradual progressive evolution to fewer and larger ears at a node until the achievement of modern high-yielding Corn Belt cultivars in the United States, which usually have one ear at a node and average less than two ears per stem.

Evolutionary changes in sorghum and pearl millet are very similar. Wild sorghum tillers well and although panicles are terminal, the stems are often branched. The most derived of the modern sorghums are likely to have a single stem with a large terminal inflorescence. The contrast between the open, lacy panicle of wild sorghum and the heavy, compact, high-yielding ones of modern cultivars is striking. The evolution of pearl millet seed heads from the numerous, small (10 cm or less) false spikes of *Cenchrus americanus* (L.) Morrone to the most-derived cultivars (over two meters in length in some cases) is nothing short of spectacular. The process, however, is the same in all these cereals.

The trend from many small inflorescences to a few or a single large inflorescence is usually accompanied by an increase in seed size. There are other selection pressures that favor large seeds, but a part of the increase may come automatically with the increased size of the inflorescence. The end product of the trend is a monstrous structure completely unadapted for survival in the wild. The head of a commercial cultivar of sunflower, an ear of maize, or an inflorescence of modern grain sorghum or grain-type pearl millet are each amazingly different from their wild progenitor forms, yet the evolutionary pathway is essentially the same in all.

Maturity is often closely controlled by daylength sensitivity. Tropical cereals like rice, sorghum, and maize are often extremely sensitive because of strong selection pressures to mature at the end of the rainy season. When these crops are moved into more temperate regions, adaptation involves a decrease in sensitivity to short days. Maturation over a long period of time has a selective value for wild plants but is detrimental to cultivated races. The disruptive selection for this trait reinforces that established by the shattering versus nonshattering characteristics.

The small grains all have terminal inflorescences, and there is no sunflower effect. An increase in uniformity of maturation is obtained by tilling over a short period of time. The life cycle of cultivars is more rigidly controlled over time than that of wild races. At a given stage, tillering essentially ceases and there is a tendency for the whole plant to mature at once. This response may also be strongly conditioned by daylength. Regardless of the mechanisms involved, selection pressures are strong in the direction of uniform maturity and maximum seed recovery at harvest.

Perennial wild grasses are notoriously poor or erratic seed setters. For them, survival does not depend on annual seed production and a full set of seed is an uncommon event. Wild annual grasses, however, set seed much better and would not survive without rather abundant yearly seed production. This phenomenon has only peripheral interest to the cereals under consideration. Rye was initially thought to have come from a perennial ancestor, but evidence shows it is from an annual progenitor. Cultivated rye was selected out of a weed rye, which in turn was derived from the wild annual progenitor *Secale cereale* L. subsp. *vavilovii* and related species in genus *Secale* (Schreiber et al., 2019). Thus, only rice may have come from perennial progenitors. In Africa, it is clear that African rice was domesticated from the wild annual *barthii* subspecies. In Asia, the picture is not as clear regarding a single or multiple origin of the two subspecies of rice, *indica* and *japonica*. Molina et al. (2011) concluded that both subspecies originated from the wild rice species *Oryza rufipogon* Griff., but Rawal et al. (2018) concluded that cultivated rice may have had three progenitors including *O. barthii, O. rufipogon* and *O. nivara* S.D. Sharma and Shastry. Only *O. nivara* is annual.

In the course of evolution in the Gramineae, it is not uncommon for florets of a spikelet to become sterile and/or reduced and for spikelets to become male, neuter, reduced, or entirely suppressed. Such reduction series are to be found in most tribes of the family. It is not uncommon under domestication for that which has been suppressed to be restored.

In *Hordeum*, there are three spikelets at a node; the central one female fertile, the lateral ones male or neuter. Throughout the entire genus, the wild barleys are two-rowed. In the cultivated six-rowed barley, the lateral spikelets are fertile. The change is rather simple genetically in that a single recessive mutation is adequate to transform a two-rowed barley into a six-rowed one. There are at least two loci involved with an allelic series at each controlling various intermediate forms (Harlan, 1968).

In sorghum, there are cultivars with two seeds per spikelet. A suppressed floret has become fertile. These twin-seeded lines are seldom very productive and have not attracted much attention. The inheritance of the character appears to be controlled by a single dominant gene.

In wheat, the number of fertile florets varies greatly and is under genetic control, although there also may be environmental influences. Some genes cause the lower florets to become sterile and the upper ones fertile; other genes reverse the arrangement. Of particular interest is the fact that even the lower florets that are "always" sterile can be made to bear seeds under certain genetic circumstances. Four doses of the Q gene are required, but it is evident that organs once suppressed can be restored and are under genetic control (Frankel et al., 1969; Wright, 1958).

In the female flowers of wild maize (teosinte), internodes are inverted to form cupules. Two spikelets (kernel-bearing structures) form within each cupule but only one develops, leading to one kernel in each hard teosinte fruit case and usually to a total of 10 or 12 kernels per ear. In the male inflorescence both spikelets produce anthers, which is also seen in maize. However, an early step in maize evolution was the restoration of the one spikelet in the female flower to fertility, leading to twice as many rows of kernels. While the final number of rows vary between maize races, most modern cultivars have 16 rows and hundreds of kernels per ear. In all maize, the restoration to fertility of the sterile floret is due to as few as three major genes, and several more of smaller effect (Doebley, 2004).

Genes for restoration of fertility to reduced spikelets and/or florets can also be found in nature. *Dichanthium* is a genus in the Andropogoneae tribe of the grass family and has paired spikelets as in maize. In the Australian species of *Dichanthium*, both members of the pair are female fertile. Crossing studies with the sibling species *Dichanthium annulatum* (Forssk.) Stapf, showed that the character was controlled by a single gene (Borgaonker et al., 1962). A gene has been found in wild populations of *Tripsacum dactyloides* L., a relative of maize, which restores fertility to both reduced spikelets and reduced florets in the female section of the inflorescence and produces perfect flowers in the male section (Dewald et al., 1987).

Increase in size of the seed head or ear is closely related to the reduction in number of inflorescences, but selection in this direction is reinforced by selection for increased seed production. The plant that contributes the most seed to a harvest is likely to contribute more offspring to the next generation. Selection is automatic in this respect, but is likely to be still further reinforced by human selection for apparent yield. All of these selection pressures go in the same direction and away from wild-type progenitors. Maize, sorghum, and pearl millet are more affected than the other cereals.

In wheat, barley, rye, oats, and rice, increased yield is achieved through an increase in tillering. Inflorescence size may also increase under selection pressure for greater seed production, but not as spectacularly as in maize, sorghum, and pearl millet.

Selection Associated with Seedling Competition

The cultivated field is a very different environment from a wild habitat. The seedbed is favorable for germination and competition with other species is reduced. The competition between seedlings of the same species, however, can become extremely intense. The first seeds to sprout and the most vigorous seedling are more likely to contribute to the next generation than the slow or weak seedlings. Within species, large seeds produce more vigorous seedlings than small seeds (Kneebone & Cremer, 1955).

Selection for highly competitive seedlings results automatically in selection for larger seeds, but the plant that produces the greatest number of seeds also has an advantage and this factor may not be compatible with the production of large seeds. Eventually, a balance is reached in which selection is continuous for a large number of seeds yielding competitive seedlings.

This balance can be easily changed by human activities. Deep planting, for example, may result in strong selection pressure for large seeds, while shallow planting may have little effect. Great variation in seed size may, therefore, be expected in cultivated cereals, but the seeds are usually (not always) larger than those of the wild races.

A trend toward lower protein and higher carbohydrate content of cereals is automatic in that most of the increase is seed size is due to an increase in endosperm. The embryo is richer in protein and oil, but does not increase in the same proportion as the endosperm. This type of selection results in increased seedling vigor.

Most wild grasses have some sort of seed dormancy, which breaks down with time. The dormancy prevents premature germination and, when it lasts for several years, helps to maintain a seed supply in the soil. There is an obvious selective advantage for this adaptation in wild plants. Wild oats, einkorn, and emmer in the Near East have an elegant adaptation to the erratic rainfall of the region. In all three wild grasses, there are two seeds in each spikelet, one without appreciable dormancy and the other sufficiently dormant that it will not normally sprout for a year or more after shedding. The nondormant seed of the pair is usually about twice as large as the dormant one. The nondormant seeds germinate with the first rains in the fall and must compete with dense populations of other annual plants. Large seed has a selective advantage under these conditions and some races of wild barley, emmer, and oats have seeds larger than many cultivated races. If the rains fail after this first emergence and the plants die without producing seed, there is still a reserve of dormant seeds that can sprout the following year.

While dormancy has adaptive value for wild and weed races, it is nonadaptive in cultivated races unless it is of short duration. Some cultivars lose dormancy altogether, but in regions where rainfall at harvest time is likely, seeds may sprout in the ear and are lost. A dormancy that breaks down between harvest and planting time may have selective value under some conditions. Automatic selection pressures are very strong for seeds that come up when planted; therefore, dormancy is much reduced in cultivated races. In a study with spontaneous finger millet, Hilu and deWet (1980) found that dormancy was reduced markedly in only four generations of selection.

Man has, no doubt, selected intensively for less chaff, but there are certain automatic selection pressures set up in the same direction. Dormancy is often controlled by inhibitors in the enclosing glumes, lemmas, and paleas. Selection for reduced dormancy may act to reduce these structures. An increase in seed size also has the feature of reducing the relative amount of chaff in the material harvested. The various selection pressures are all inter-locking and trending in the same directions.

Selection Associated with Crop-Weed Interaction

When man tills the soil and prepares a seedbed for his crops, he also provides an environment favorable for wild species with adaptations that can take advantage of the new situation. The cereals have responded with weed races adapted to the conditions of cultivation. The morphology and adaptations of the weed races are generally intermediate between the wild and cultivated races. The weeds are adapted to disturbed environments, but retain the shattering habit and, frequently, the dormancy and seed appendages of the wild races.

There is good evidence that the companion weeds of cultivated plants have played important roles in crop evolution. Wherever the crop and weed occur together on a massive scale, hybrid swarms and evidence of introgression can be found. The barriers to gene flow are always rather strong, and hybridization does not occur on a massive scale. One popula-tion is not going to be swallowed up by the other. Instead, we have two separate populations growing side by side and maintaining their own heredities, but occasionally they do cross and germplasm is exchanged. A differentiation–hybridization cycle is set up and potential variability is released.

In crosses between wild and domesticated races of cereals, either mor-phological type can be quickly recovered in backcrosses. The small, two-ranked, fragile ear of wild maize is strikingly different in appearance from the large, multi-rowed, nonfragile ear of cultivated maize, yet both morpho-logical types can be recovered in a single F_2 population (George Beadle, per-sonal communication; Galinat, 1971): either only a few genes are involved or the genes that are involved are tightly linked on only a few chromosomes. Similar crosses in other cereals give similar results, which is expected con-sidering the observed interaction between spontaneous and cultivated races in the field.

Although gene flow can be easily detected as we have shown by the mimicry of weed races and the infusion of the cultivated traits into them, intermediate morphologies are rare. Fragile six-rowed barleys are encountered as a byproduct of spontaneous × cultivated crosses in the Near East (Zohary, 1964, 1971), but are nonadapted and quickly disappear from hybrid swarm populations. I have collected both shattering caudatums and shattering guinea sorghums in Africa, but plants of this morphology are rare and probably ephemeral as well. Even in interbreeding populations, the morphologies are either spontaneous or cultivated.

The system is a remarkably elegant evolutionary adaptation. Too much crossing would degrade the crop, and the weed and cultivated races would merge into one population, possibly resulting in abandonment of the crop. Too little crossing would be ineffective. The barriers to gene flow must be strong but incomplete for the system to work, and the fact that crop–weed pairs have evolved in so many crops is an indication that natural selection has operated to adjust the amount and frequency of hybridization to a range somewhere near the optimum evolutionary efficiency.

Differentiation–Hybridization Cycles

Cultivated plants have the capacity to evolve rapidly. Rapid bursts of evolution are possible only when sufficient diversity exists to allow for great change in a short period of time. The ultimate source of variability may be mutation, but individual mutations in one population will take much too long to accumulate to allow for this level of diversity to build up. Thus, mutations must occur in many populations over a long period of time and then exchange alleles by mixing. This is a variation on the theme of the differentiation–hybridization cycle in which variability already accumulated can be exploited.

The crop–weed interaction is only one system by which differentiation–hybridization cycles can be set up in cultivated plants. Another is more or less automatically built into the traditional agricultural system. Farmers are basically sedentary. They settle down in an area and occupy it for long periods of time. This results in an array of varieties adapted to that particular area. Occasionally, farmers move, taking their seed stocks or other planting materials with them. In the new location there is an opportunity for the transported varieties to cross with the local types. Populations separated geographically and differentiated ecologically are thus brought together, crosses occur, and a cycle is completed. The past movement of certain races of maize can be traced today from Mexico to South America, where the races were modified by introgression with South American races; then this material returned at a later date to Mexico where it introgressed with maize and formed new hybrid races with Mexican races (Bedoya et al., 2017; Wellhausen et al., 1952). An analysis of the great diversity in Turkish popcorns by Anderson and Brown (1953) indicated that the irruption of variability could be traced to two different races introduced into Turkey by different routes. The same authors presented evidence to the effect that Corn Belt maize in the United States is derived from the interaction of northern flints and southern dents brought together unintentionally by white settlers moving into the US Corn Belt. The migration of identifiable races of American cottons and their periodic introgression also have been documented by Hutchinson (1959).

Patterns of this nature can be traced in a number of crops. Populations of cultivated plants are far more mobile than populations of wild species because they are transported by man and go with him on his wanderings over the face of the earth. This can readily result in the separation of plant populations by geographic isolation and the breakdown of isolation by bringing populations together again. There can be little doubt that these movements and migrations have profoundly affected the evolution of cultivated plants by exposing populations to infusions of germplasm from other domesticated races as well as from various wild relatives of the crop.

Cycles may also be set up by cultural practices. In many parts of West Africa, sorghum is transplanted like rice in Asia. Seedlings are grown in a sandy seedbed, pulled up by the roots, and planted in deep dibble holes in the field. This is done as waters recede after the annual flood of a river or as wet areas dry up at the end of the rains. These cultivars must mature seed on residual moisture only and must be ephemeral, short-season types. The same cultivator may grow full-season types during the rainy season. The transplanted race may mature in 90 days; the rainfed one may take 180 days. Such practices set up separate populations that have little chance of interacting with each other; one population matures while the other is in the seedling stage.

The most common rainfed race of sorghum is guinea and the most common transplant race is durra. We have, however, detected some guinea–durra interaction (Harlan & Stemler, 1976). This can come about by late summer or fall planting of a mixture of the two races when the shortening days may bring them into bloom together. Mixtures of this kind may occur through careless handling by the cultivator, but are more likely through seed purchase at the local market.

Thus, populations are fragmented by human activity and often kept apart by agricultural practice. Such barriers leak, and differentiation–hybridization cycles are set up that enhance variability and broaden the base of plant selection.

For differentiation to take place, populations must be fragmented in some way and kept apart genetically. Several isolating mechanisms among cereals are well known, for example, geographic and ecological separation, differences in time of blooming, self-fertilization (barley, wheat, oats, rice, and sorghum), translocation races (rye), polyploid races (wheat and oats), gametophytic factors (maize), cryptic chromosomal differences (rice), and meiotic irregularities (wheat). No one scheme is necessarily better than another. Any barrier to gene flow will permit populations to fragment and accumulate genetic differences among the subpopulations. Sometimes a combination of several mechanisms can be demonstrated. The only qualification is that if the differentiation–hybridization cycle is to function, the isolation cannot be absolute or permanent. Sooner or later, the separated populations must be brought together again to permit some hybridization.

The appropriate degree of differentiation depends on the amount of buffering in the genetic system. By genetic buffering, we mean essentially the amount of redundancy of genetic information. A self-fertilizing diploid would be presumed to be weakly buffered. Crossing between cultivars should result in a rather major release of potential variability; the variability should be largely oligogenic in which relatively few genes have conspicuous effects and truly wide hybridizations should be disastrous, if at all possible. The system is rather well illustrated by Figure 6.1.

A cross-fertilizing diploid such as maize should be somewhat better buffered by carrying many genes in a heterozygous condition. Narrow crosses should have less obvious effects and wider crosses should be tolerated. Variation should still tend toward the oligogenic, at least when compared to a polyploid like wheat. Hexaploid wheat is better buffered than tetraploid wheat since the redundancy in genetic information is greater. The elaborate and elegant chromosome engineering of E.R. Sears (1969) could not have been performed among tetraploids, for example. At the hexaploid level, narrow crosses have relatively little effect and decidedly wide crosses are tolerated (e.g., hexaploid wheat with *Aegilops, Secale,* and *Agropyron*).

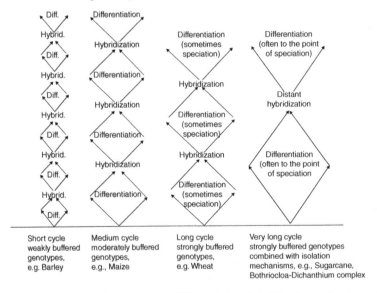

Figure 6.1 Schematic diagram of differentiation–hybridization cycles depending upon the degree of buffering of genotypes in cultivated plants. *Source:* Harlan (1966). Reproduced with permission from John Wiley & Sons.

Finally, the most highly buffered system in the scheme is the case of a high polyploid propagated vegetatively so as to escape the penalties of sterility. Such systems can withstand the shock of distinctly alien germplasm and the widest crosses are thereby tolerated.

Although there are, of course, many gradations between the illustrations given above, all of the systems work. Barley with its short cycle was probably the most important cereal crop on earth at one time. It is still important and conspicuously variable, and has a rather high yield potential. The variability may be somewhat deceiving because of the oligogenic nature, but there seems to be no great advantage or disadvantage to either a long or a short differentiation–hybridization cycle. In each case, the variability accumulated during the differentiation phase is exploited by hybridization at the appropriate stage of differentiation. Natural selection appears to have operated to adjust the length of the cycle to the degree of buffering. If the cycle is too short, there is little effect; if the cycle is too long, it cannot be completed.

Other Selection Pressures

Throughout the process of domestication, deliberate human selections have been superimposed on automatic selection pressures. In many cases, they are in the same direction and reinforce each other. In selecting for apparent yield, man also will select for larger inflorescences, larger seeds, more seeds, better seed set, more determinate growth, daylength sensitivity, easier threshing, and so on. Deliberate selection adds new dimensions to the process, for human selection may be more intense and absolute and is often biologically capricious or even whimsical.

Without deliberate selection, a given genotype may have a low statistical chance of contributing offspring to the next generation. It is a component of a population and even if it is not among the best fitted to the environment, elimination may be relatively slow. However, most cultivators in what we call "primitive agriculture" are very particular about the seed they sow. Each year at harvest time, they carefully choose certain heads of sorghum or ears of maize and seed from these *only* will be planted for the next generation. To be sure, this procedure is less universal among the small grains, but individual plant selection is still common.

This practice provides a new order to selection pressure. The population becomes an array of deliberately chosen components. It may still be rich in variation because cultivators of traditional agriculture have an appreciation for mixtures, but the mixtures will conform to whatever an individual selector chooses. The total potential range of variation will be fragmented into landrace populations or primitive cultivars. Different cultivars will be grown for different purposes or to fit different ecological niches of the agricultural system.

Man selects for color, flavor, texture, and storage quality. He selects maize for popping, boiling, eating off the cob, flour quality for making hominy, and for ceremonial purposes in religious rites. He selects sweet sorghum for chewing, white-seeded types for bread, small dark red-seeded types for beer, and strong-stemmed, fibrous types for house construction and basketry. He selects glutinous rice and nonglutinous rice, long-grained rice and short-grained rice, red rice, white rice, and aromatic rice. He selects barley for food, barley for beer, and barley for livestock feed. He selects grains that grind well or that process well in a mortar. Man delights in bright colors and in curious and unusual variants, and he may select for several different types. High yield is seldom a factor in traditional agriculture, but consistent and reliable yield is absolutely essential; man knows his materials well because survival depends on it.

Man in traditional agriculture has an intuitive feeling for nutritive value. Certain cultivars are said to be good for pregnant women, others for nursing mothers. Many cultivars are prized as food for young children and some are said to be "strong" and reserved for periods of heavy work in the field. Without chemical analyses or laboratory rats for testing, the intuitive feeling usually has considerable nutritional merit. Under automatic selection, the fact that a line of sorghum makes good dumplings confers on it no particular fitness for survival, but under human selection, fitness may be total. The line survives only because man plants it.

Other Seed Crops

The principles described for cereals apply in general to other seed crops. The differences are largely in specific details. In leguminous and cruciferous seed crops, for example, indehiscent pods and siliques evolve instead of nonfragile rachises. In the pulses, this may be a rather major change.

A legume is a one-celled pod with two sutures. In many legumes, the pod has a specialized inner layer that contracts at maturity. When the pressure is sufficient and the sutures sufficiently weakened, the legume opens explosively. The inner tissues contract and the halves of the pod roll up into a spiral. The pop is clearly audible even on small pods and the seeds are often snapped several meters from the plant. The legumes of some large tropical tree species pop open with a sound like a small firecracker and seeds are disseminated with considerable force.

The inner layer that operates the dispersal mechanism may be suppressed in cultivated races of legumes. This is the most diagnostic difference between wild and domestic races and has been used to identify cultivated beans and lima beans from Guitarrero Cave in Peru dated to over 10,000 BP (Kaplan et al., 1973;

Lynch, 1980). Where seeds are contained in pods and capsules, nonshattering evolves through indehiscence in a way analogous to nonshattering in cereals, although different tissues are involved.

Other evolutionary trends are also similar: we may expect larger seed, loss of dormancy, more determinate growth, more flowers, and larger inflorescences. They are similar because the selection pressures are similar. Weed races are likely to appear as well. Some crops may be less disposed to the evolution of weeds than are the cereals, but among annuals there are few exceptions.

There is no space here to go into characteristics of population dynamics, but the reader may wish to look into some of the literature. Brown et al. (1990) and Larson et al. (2014) are good places to start. Many studies, both experimental and theoretical, have been published dealing with modes of gene interaction, selection for fitness, disruptive selection, epistatic selection, linkage, linkage disequilibrium, maintenance of polymorphisms, and so on. Through molecular marker analyses, such studies can be conducted on wild species and even on trees whose genetics are otherwise unknown. An early, valuable example of such studies are the composite cross (CC) populations of barley. CC II was generated by my father (H. V. Harlan) in 1928, who crossed 28 carefully chosen parents in all possible combinations (378 crosses). This mixture had been advanced over generations since the population was created at Davis, CA, and has been studied extensively because changes can be monitored over time using seed saved from previous generations. Other composite cross populations have been generated using other barley parents, even one using over 6000 entries, and the several CCs can be compared over time. Other species, including outbreeders, have also been analyzed for population structure in similar fashion. Generations are advanced without deliberate selection, allowing automatic selection forces of farming practices and environment to act on the populations.

High variability has been maintained in CC II despite 90 years of selfing. All of the CCs increased in yield without deliberate selection for yield. Under cultivation, there is automatic selection for higher yield, faster in early generations, but still continuing. The increase was largely due to more seeds per plant rather than seed size. Ears became heavier and more compact. Certain multilocus associations appeared at high frequencies even when the genes are not linked (linkage disequilibrium). These associations appeared in early generations and are adaptive. The patterns that developed at Davis, CA, with a Mediterranean climatic regime were different from those developed in the same populations grown in areas of continental climate. Linkage disequilibrium is stronger in self-fertilizing species than in outbreeders, and inbreeding populations are more structured. Alleles that confer disease resistance, generally, have negative effects on yield in the absence of the disease and sometimes even in its presence. There is some cost in resistance mechanisms (Allard, 1990).

New population genomics tools (Tian et al., 2009) combined with different population designs (e.g., Bandillo et al., 2013; McMullen et al., 2009) are allowing the close inspection of genomic regions on which natural or artificial selection has acted, expanding the range of knowledge and depth of coverage of crop evolution. A recent comprehensive molecular study on barley nicely complements the older studies on the CC populations (Milner et al., 2019). What we are seeing in these studies is an increase in fitness or adaptation over the years to the environment and farming practices and local diseases in each region. Most of the loci studied had something to do with fitness. The loci might control conspicuous morphological traits, disease reaction, molecular variants, and so on, but all of them were related in some fashion to reproductive capacity. It seems that farming *is* plant breeding, and these studies give us some insight into landrace formation in traditional agriculture and how wild plants might adjust to cultivation.

Despite the spectacular arrays of variation in cultivated plants, the genetic differences between wild and cultivated races do not appear to be enormous. Biological speciation hardly ever occurs and unless polyploidy is involved, strong genetic barriers have rarely developed. The number of genes controlling wild versus cultivated morphologies is often rather small, and these are likely to be linked into a few linkage blocks.

At this point I would like to mention some personal experiences with attempting to domesticate native American grasses. It was intended that these would be used for forage and revegetation of abandoned farmlands. The species chosen were climax perennials, not weedy, and with inherently poor seed production characteristics. Even though forage production was the ultimate use, seed and seedling characteristics were the most critical factors in domestication. Seed set, seed retention, seed size, seedling vigor, and stand establishment were critical in developing something useful. Sources of wild populations were from the southern Great Plains of the United States. In most species, substantial progress was made in reducing shattering, increasing seed production, and developing larger seed and more seedling vigor. Several species were capable of yielding in the range of 600–700 kg/ha, which is within the range of cereals such as wheat, rice, and maize, under conditions of subsistence agriculture with depleted soils and no fertilizer supplements.

Forage crop breeding, in general, is relatively new and many of the materials in use are simply selected from naturally occurring, adapted populations. The ladino-type white clover, the southern smooth bromegrasses, and the many landraces of alfalfa, such as "Flamande," "Ladak," "Semipalatinsk,"

"Indian," and California, Oklahoma, and Kansas commons, are examples. In weeping lovegrass, standard, "Ermelo," and "Morpa" strains are introduced populations from different parts of Africa. "Greenfield" bermudagrass is a selection from spontaneous populations and "Coastal" was a chance nursery hybrid. Pangolagrass is apparently a naturally occurring sterile hybrid. Numerous other examples could be cited. The point is that there is a great deal of variation in natural populations available for immediate exploitation. Domestication is not as difficult as one might think.

Domestication of Vegetatively Reproduced Crops

Vegetative Propagation

Among root crops, yams (*Dioscorea* spp.) can be taken as a model because so many species have been cultivated over widely separated parts of the world. Table 6.2 shows that man has had a rather extraordinary interest in digging tubers of *Dioscorea* out of the ground. Between 50 and 100 species have probably been used for human food, although many of them must be detoxified to make them safe to eat.

Yams have uses other than food and we have no way of knowing what it was that first attracted man to these plants. The vines are used for cordage. Many of the African and Asian species contain water-soluble alkaloids that can be extracted and used as poisons against fish, monkeys, insects, tigers, and humans. Steroidal poisons occur in yams of both the Old World and the New World and have been used for arrow and fish poisons and against lice (Coursey, 1972). In any case, the vines and leaves of *Dioscorea* are easily recognized and the tuber is readily located at the base of the vine. Preagricultural people knew about them wherever they occurred in the wet tropics.

The species that have attracted man are primarily adapted to savanna zones with a pronounced wet and dry season. The tuber is a storage organ adapted to such a regime. As pointed out in Chapter 2, the tubers sprout at the onset of the rains and the vines grow with remarkable vigor; virtually the entire contents of the tuber are mobilized and translocated upward. At the end of the rains, the process is reversed and nearly all of the food stored in the vines is translocated downward and the tuber grows very rapidly. The vine dies and the tuber remains dormant through the dry season. The life cycle is geared toward survival in climates with sharply contrasting wet and dry seasons.

This poses a problem of exploitation. Premature digging will kill the plant and net very little. The tuber is formed in a rush at the onset of the dry season. This is why the Aborigines made use of "calendar" plants to signal the digging

Table 6.2   The more important species of *Dioscorea* used for human food (adapted from Coursey, 1972).

Asia	Africa	America	Australia
	Major species		
D. alata L.	*D. rotundata* Poir. (syn. *D. cayenesis* subsp. *rotundata*)	*D. trifida* L.f.	
	D. bulbifera L.		
D. esculenta (Lour.) Burkill	*D. cayenensis* Lam.		*D. hastifolia* Nees
	Secondary species		
D. bulbifera L.	*D. dumetorum* (Knuth.) Pax.	*D. altissima* Lam.	*D. transversa* R. Br.
D. pentaphylla L.	*D. preussii* Pax		
D. nummularia Lam.	*D. praehensilis* Benth.		
D. oppositifolia L.	*D. sansibarensis* Pax.		
D. japonica Thunb.	*D. colocasiaeifolia* Pax.		

season. We have mentioned previously the religious sanctions of yam eaters in Australia, the Andaman Islands, and West Africa. People learned long ago that premature harvesting is disastrous to the crop, and this knowledge was woven into their religious systems. We have also called attention to the practice of planting the head back at harvest time, which existed even among Australian Aborigines and Andamanese who otherwise practice no horticulture.

A practice of "protoculture" was described by Chevalier (1936) in the Ubangui–Chari region of equatorial Africa. Here the people were harvesting *Dioscorea dumetorum* (Knuth.) Pax in the wild. They used what they needed immediately and planted the surplus near camp for future use. In other cases it has been observed that wild yams are brought to camp where the woody heads are cut off and discarded. These may sprout at the proper time of year. The step from precultural and protocultural practices to fully cultural practices is a small one. In the case of yams, it merely meant the storage of tubers from harvest time to planting time, and some kinds of yams store very well.

The domestication of manioc was perhaps even easier. *Manihot* is better suited to the wet tropics, although it is also very drought resistant and has a wide ecological amplitude. It can be reproduced by stem cuttings. All that is necessary is to cut off a branch and stick it in the ground during the rainy season and tubers will be produced. More important, a plant produces a number of tubers and harvesting can be done all year long. There is no need to protect the plant for part of the year.

Among vegetatively propagated plants, selection is absolute and the effects immediate. If clones should be found that are better tasting, less poisonous, more poisonous, more productive, and so on, they can be propagated and cultivars are developed immediately. In both yams and manioc, many clones have lost the power to reproduce sexually. They may not bloom at all, or the flowers may be deformed and sterile. These clones are fully domesticated and entirely dependent on man for survival. In manioc, this seems to occur more frequently in the most poisonous types, suggesting that man has actually selected for increased prussic acid. In the wet tropics especially, this affords some protection against insects and wild mammals. Methods of detoxification are laborious but well worked out.

Loss of sexual reproduction is well known in the triploid and tetraploid bananas and plantains. Presumably, hybrids between *Musa acuminata* Colla and *M. balbisiana* Colla occurred naturally through the function of unreduced eggs, resulting in sterile triploids. Undoubtedly, this happened a number of times to the delight of tropical agriculturalists of Southeast Asia and neighboring islands because the diploid bananas are often very seedy, and banana seeds can crack your teeth. Sterile diploid mutations have also occurred and have been propagated vegetatively.

The ease with which vegetatively propagated plants can be brought under domestication is one argument advanced in support of the idea that tropical "vegeculture" is older than seed agriculture. There can be no question of its simplicity nor of its potential for instant domestication. Firm evidence of its greater antiquity remains to be developed, but there is some logic in the argument.

Grafting

The invention of grafting techniques has extended the advantages of vegetative propagation to many crops otherwise propagated by seed. Some of these have been domesticated in recent times and we have full records of the process. The American grape (*Vitis* spp.) is one example. No doubt, the Native Americans knew of some of the more fruitful and palatable vines, but I know of no evidence that they ever tried to cultivate them. Early selections by Europeans were all from wild and weedy populations. To be sure, the behavior of European settlers had some influence on native populations of grapes.

The fox grape (*Vitis labrusca* L.) is well adapted to the edges of woods, fence rows, and so on, where birds disseminate the seeds. When settlers began to make farms out of forests, the forest margin habitat was increased enormously under conditions that favored hybridization. It is from these variable populations that some of our better-known American grapes were selected. The "Concord" cultivar, perhaps the most popular of all, was selected by E.W. Bull, of Concord, MA, from his pasture. The cultivar "Catawba" was selected by Major Hadley in Georgetown, District of Columbia. "Isabella," "Rebecca," "Niagara," and many other well-known cultivars were selected over the decades. A work by Charles Downing (1869) lists 144 selected varieties of American grapes. Almost all of these came from wild and weedy populations and only a few from artificial breeding attempts. Several different American species contributed cultivars. Repeated attempts were made to introduce European grape (*V. vinifera* L.) germplasm by selection and by crossing with American species. The early attempts were largely failures, but large quantities of European vine cuttings were taken to California in the 1850s and 1860s; and direct selection of spontaneous seedlings seemed to give the best cultivars. Today, European grapes are preferred for wine and are grown in several states (Krochmal & Grierson, 1961).

Most improved clones of pecan [*Carya illinoensis* (Wangenh.) K. Koch] were simple selections from native bottomland stands of wild trees. Many individual trees contributed to the class called "paper shelled." Pecan fanciers and nurserymen searched through the natural stands and when they found a tree bearing nuts to their liking (thin-shelled, good flavor, large size, etc.), they simply cut off some bud-wood and grafted it onto seedling stocks. This again, is instant domestication. Similar procedures have led to cultivars of hickories, wild plum, mulberry, hazelnut, and many others.

Ornamentals

Finally, there is a large class of recent domesticates consisting of ornamentals. Many of these are perennials and can be propagated vegetatively. This means that the sterile hybrids of wide crosses can be used, and some of our most striking garden flowers have such complex origins. Wild forms are much in demand for breeding stock, and the most eloquent testimony to this is the scandalous smuggling of wild tulips (*Tulipa* spp.) from Turkey. Wild tulips are protected by law, but large numbers of bulbs are dug up yearly and are smuggled out of the country.

Summary

These examples demonstrate not only the ease with which vegetatively propagated crops can be domesticated, but also the great variability of natural or spontaneous populations. Strikingly superior types can be found by screening large natural populations. Furthermore, selections that would perish under natural conditions can be salvaged for later use. The navel orange, for example, is seedless and could not have reproduced itself. Seedless grapes are in the same class, to say nothing of the thousands of sterile clones of bananas, yams, manioc, and other crops already mentioned.

Conclusions

Heraclitus was right: the joining together of unlikes and the operation of "strife" or selection pressures have resulted in the production of domesticated races of plants. Natural populations of many species are variable and one can go a long way toward domestication simply by selecting and propagating desirable types that are already available. In seed crops, *planting* is the key operation. This practice alone sets up an entire syndrome of interlocking automatic selection pressures. Superimposed on the automatic pressures are those of deliberate selection; these are extremely powerful, very artificial, and often capricious. The result is the preservation of types that are entirely incapable of survival without the care of man.

Human activities, often unintentionally, established fairly effective plant breeding systems. Migration, trade, cultural practices, or even deliberate manipulation have brought about repeated stirring of the gene pools and the development of differentiation–hybridization cycles.

These dynamics have resulted in great morphological changes without substantial change in the genetic background; speciation rarely occurs under domestication. The parts of the plant that show the greatest morphological alterations are the parts most valued by man. (This was noted by both Darwin (1859) and de Candolle (1885).) Mutations and gene combinations that cause striking morphological modifications are relatively common, but man must intervene and propagate them or they will be pruned out by natural selection. Under domestication, modification can build on modification until the end products are radically different in appearance from their wild progenitors.

References

Allard, R. W. (1990). Future directions in plant population genetics, evolution, and breeding. In A. H. D. Brown, M. T. Clegg, A. L. Kahler, & B. S. Weir (Eds.), *Plant population genetics, breeding, and genetic resources* (pp. 1–20). Sunderland, MA: Sinauer Associates Inc.

Anderson, E., & Brown, W. L. (1953). The popcorns of Turkey. *Annals of the Missouri Botanical Garden, 40*, 33–49. https://doi.org/10.2307/2398973

Bandillo, N., Raghavan, C., Muyco, P., Sevilla, M., Lobina, I., Dilla-Ermita, C., . . . Mauleon, R. (2013). Multi-parent advanced generation inter-cross (MAGIC) populations in rice: Progress and potential for genetics research and breeding. *Rice (N. Y.), 6*(1), 11. https://doi.org/10.1186/1939-8433-6-11

Bedoya, C. A., Dreisigacker, S., Hearne, S., Franco, J., Mir, C., Prasanna, B. M., . . . Warburton, M. L. (2017). Genetic diversity and population structure of native maize populations in Latin America and the Caribbean. *PLoS One, 12*(4), E0173488. https://doi.org/10.1371/journal.pone.0173488

Borgaonker, D. S., Harlan, J. R., & deWet, J. M. J. (1962). A cytogenetical study of hybrids between *Dichanthium annulatum* and *D. fecundum*. *Proceedings of the Oklahoma Academy of Science, 42*, 13–16.

Brown, A. H. D., Clegg, M. T., Kahler, A. L., & Weir, B. S. (Eds.) (1990). *Plant population genetics, breeding, and genetic resources*. Sunderland, MA: Sinauer Associates Inc.

Chevalier, A. (1936). Contribution a l'étude de quelques espèces africaines du genre *Dioscorea*. *Bulletin du Muséum National d'Histoire Naturelle, 8*, 520–551.

Coursey, D. G. (1972). The civilizations of the yam: Interrelationships of man and yams in Africa and the Indo-Pacific region. *Archaeology & Physical Anthropology in Oceania, 7*, 215–233.

Darwin, C. R. (1859). *On the origin of species by means of natural selection*. London: J. Murray.

de Candolle, A. (1885). *Origin of cultivated plants*. New York, NY: D. Appleton and Company. https://doi.org/10.5962/bhl.title.29067

Dewald, C. L., Burson, B. L., deWet, J. M. J., & Harlan, J. R. (1987). Morphology, inheritance, and evolutionary significance of sex reversal in *Tripsacum dactyloides (Poacaae)*. *American Journal of Botany, 74*, 1055–1059. https://doi.org/10.1002/j.1537-2197.1987.tb08715.x

Doebley, J. (2004). The genetics of maize evolution. *Annual Review of Genetics, 38*, 37–59. https://doi.org/10.1146/annurev.genet.38.072902.092425

Downing, C. (1869). *The fruits and fruit-trees of America*. New York, NY: John Wiley & Sons.

Frankel, O. H., Shineberg, B., & Munday, A. (1969). The genetic basis of an invariant character in wheat. *Heredity, 24*, 571–591. https://doi.org/10.1038/hdy.1969.79

Galinat, W. C. (1971). The origin of maize. *Annual Review of Genetics, 5*, 447–478. https://doi.org/10.1146/annurev.ge.05.120171.002311

Harlan, J. R. (1966). Plant introduction and biosystematics. In K. J. Frey (Ed.), *Plant Breeding* (pp. 55–83). Ames, Iowa: Iowa State University Press.

Harlan, J. R. (1968). On the origins of barley. In J. R. Harlan (Ed.), *Barley: Origin, botany, culture, winterhardiness, genetics, utilization, USDA Handbook no. 338*. Washington, DC: U.S. Government Printing Office.

Harlan, J. R., deWet, J. M. J., & Price, E. G. (1973). Comparative evolution of cereals. *Evolution, 27*, 311–325. https://doi.org/10.1111/j.1558-5646.1973. tb00676.x

Harlan, J. R., & Stemler, A. B. L. (1976). The races of sorghum in Africa. In J. R. Harlan, J. M. J. deWet, & A. B. L. Stemler (Eds.), *The origins of African plant domestication*. The Hague, The Netherlands: Mouton. https://doi. org/10.1515/9783110806373.465

Hillman, G. C., & Davies, M. S. (1990). Domestication rates in wild-type wheats and barley under primitive conditions. *Biological Journal of the Linnean Society, 39*, 39–78.

Hilu, K. W., & deWet, J. M. J. (1980). The effect of artificial selection on grain dormancy in *Eleusine* (Gramineae). *Systematic Botany, 5*, 54–60. https://doi. org/10.2307/2418735

Hutchinson, J. (1959). *The application of genetics to cotton improvement*. Cambridge, UK: Cambridge University Press.

Kaplan, L., Lynch, T. F., & Smith, C. E. (1973). Early cultivated beans (*Phaseolus vulgaris*) from an intermontane peruvian valley. *Science, 179*, 76–77. https://doi.org/10.1126/science.179.4068.76

Kneebone, W. R., & Cremer, C. L. (1955). The relationship of seed size to seedling vigor in some native grass species. *Agronomy Journal, 47*, 472–477. https://doi.org/10.2134/agronj1955.00021962004700100007x

Krochmal, A., & Grierson, W. (1961). Brief history of grape growing in the United States. *Economic Botany, 15*, 114–118. https://doi.org/10.1007/ BF02904083

Larson, G., Piperno, D. R., Allaby, R. G., Purugganan, R. G., Andersson, M. D., Arroyo-Kalin, M., . . . Fuller, D. Q. (2014). Current perspectives and the future of domestication studies. *Proceedings of the National Academy of Sciences of the United States of America, 111*, 6139–6146. https://doi. org/10.1073/pnas.1323964111

Lynch, T. F. (1980). *Guitarrero cave: Early man in the Andes*. New York, NY: Academic Press.

McMullen, M. D., Kresovich, S., Villeda, H. S., Bradbury, P., Li, H., Sun, Q., . . . Brown, P. (2009). Genetic properties of the maize nested association mapping population. *Science, 325*, 737–740. https://doi.org/10.1126/ science.1174320

Milner, S. G., Jost, M., Taketa, S., Mazón, E. R., Himmelbach, A., Oppermann, M., . . . Schüler, D. (2019). Genebank genomics highlights the diversity of a global barley collection. *Nature Genetics, 51*(2), 319–326.

Molina, J., Sikora, M., Garud, N., Flowers, J. M., Rubinstein, S., Reynolds, A., . . . Purugganan, M. D. (2011). Molecular evidence for a single evolutionary origin of domesticated rice. *Proceedings of the National Academy of Sciences of the United States of America, 108*, 8351–8356. https://doi.org/10.1073/pnas.1104686108

Patrick, G. T. W. (1889). *The fragments of Heraclitus of Ephesus.* Baltimore, MD: N. Murray. https://doi.org/10.1037/12370-000

Rawal, H. C., Amitha Mithra, S., Arora, K., Kumar, V., Goel, N., Mishra, D., . . . Solanke, A. U. (2018). Genome-wide analysis in wild and cultivated *Oryza* species reveals abundance of NBS genes in progenitors of cultivated rice. *Plant Molecular Biology Reporter, 36*(3), 373. https://doi.org/10.1007/s11105-018-1086-y

Schreiber, M., Himmelbach, A., Börner, A., & Mascher, M. (2019). Genetic diversity and relationship between domesticated rye and its wild relatives as revealed through genotyping-by-sequencing. *Evolutionary Applications, 12*, 66–77. https://doi.org/10.1111/eva.12624

Sears, E. R. (1969). Wheat cytogenetics. *Annual Review of Genetics, 3*, 451–468. https://doi.org/10.1146/annurev.ge.03.120169.002315

Tian, F., Stevens, N. M., & Buckler, E. S. (2009). Tracking footprints of maize domestication and evidence for a massive selective sweep on chromosome 10. *Proceedings of the National Academy of Sciences of the United States of America, 106*(Suppl. 1), 9979–9986. https://doi.org/10.1073/pnas.0901122106

Wellhausen, E. J., Roberts, L. M., Hernandex, E., & Mangelsdorf, P. C. (1952). *Races of maize in Mexico.* Cambridge, MA: Bussey Institution of Harvard University.

Wilke, P. J., Bettinger, R., King, T. F., & O'Connell, J. F. (1972). Harvest selection and domestication in seed plants. *Antiquity, 46*, 203–209. https://doi.org/10.1017/S0003598X0005362X

Wright, G. M. (1958). Grain in the glume of wheat. *Nature, 181*, 1812–1813. https://doi.org/10.1038/1811812a0

Zohary, D. (1964). Spontaneous brittle six-rowed barleys, their nature and origin. In Centre for Agricultural Publications and Documentation (ed.), *Barley Genetics I: Proceedings of the 1st International Barley Genetics Symposium. 26–31 August, 1963, Wageningen, The Netherlands.* Wageningen, The Netherlands: Centre for Agricultural Publications and Documentation.

Zohary, D. (1971). Origin of south-west Asiatic cereals: Wheats, barley, oats, and rye. In P. H. Davis (Ed.), *Plant life of south-west Asia* (pp. 235–263). Edinburgh, Scotland: Botanical Society of Edinburgh.

7

Space, Time, and Variation

> . . . *some species can indeed be said to have had a single and sudden origin, localized and capable of being located. With others, however, the origin is no origin at all but a gradual transformation extending over wide areas and long periods and shifting its focus in the course of time. Between the two is every gradation.*
>
> C. D. Darlington and E. K. Janaki Ammal (1945)

Kinds of Patterns of Variation

Geographic patterns of variation have historically been used to trace the origin and evolution of cultivated plants. We have seen that Vavilov (1926, 1951) thought that areas of maximum genetic diversity represented

Source: Patricia J. Scullion

Harlan's Crops and Man: People, Plants and Their Domestication, Third Edition.
H. Thomas Stalker, Marilyn L. Warburton, and Jack R. Harlan.
© 2021 American Society of Agronomy, Inc. and Crop Science Society of America, Inc.
Published 2021 by John Wiley & Sons, Inc.
doi:10.2135/harlancrops

centers of origin and that the origin of a crop could be identified by the simple procedure of analyzing variation patterns and plotting regions where diversity was concentrated. It turned out that centers of diversity are not the same as centers of origin, yet many crops do exhibit centers of diversity. The phenomenon is real and requires explanation. What causes variation to accumulate in secondary centers is not completely understood, but some observable factors are:

1) A long history of continuous cultivation.
2) Ecological diversity, many habitats accommodate many races.
3) Human diversity, different tribes are attracted to different races of the crop.
4) Introgression with wild and weedy relatives or between different races of a crop.

There may be other causes, but the reasons for secondary centers are human, environmental, and the internal biological dynamics of hybridization, segregation, and selection. A crop-by-crop analysis shows the situation to be much more complex than that conceived by Vavilov. Many crops did not originate in Vavilovian centers at all, and some do not have centers of diversity; several can be traced to very limited and specific origins, and others seem to have originated all over the geographical range of the species. It seems evident that if a crop originated in a limited area and did not spread out of it, the center of origin and the center of whatever diversity there may be would coincide. Both space and time are involved, and different crops have different evolutionary patterns. The main patterns can be classified and are described in the following sections.

Endemic

Crops that originated in a limited area and did not spread appreciably. Examples: *Urochloa deflexa* (Schumach.) H. Scholz in Guinea, *Ensete ventricosum* (Welw.) Cheesman in Ethiopia, *Digitaria iburua* in West Africa, *Setaria parviflora* in ancient Mexico (Callen, 1967), and *Panicum hirticaule* J. Presl in modern Mexico. Distributions of several endemic species in West Africa are illustrated in Figure 7.1.

Semiendemic

Crops that originated in a definable center and with limited dispersal. Examples: *Eragrostis tef* and *Guizotia abyssinica* (L.f.) Cass. are Ethiopian domesticates; both are grown on a limited scale in India. Basic to the

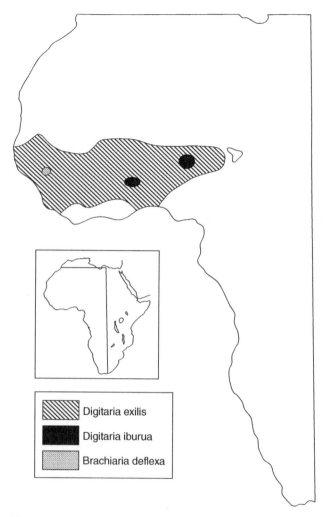

Figure 7.1 Some lesser millets of West Africa. *Brachiaria deflexa* and *Digitaria iburua* are endemic crops.

Ethiopian diet, they are not important elsewhere. African rice (*Oryza glaberrima*) is another example (Figure 7.2); the center of origin is probably the flood basin of the central Niger (Portéres, 1956) and then distributed to Senegal, southward to the Guinea coast, and eastward as far as the Lake Chad area. Some of the minor tubers of the South American highlands, such as *Oxalis tuberosa* Molina, *Ullucus tuberosus* Caldas, and *Tropaeolum tuberosum* Ruiz and Pavón, would also fall into this class (León, 1964).

Figure 7.2 Distribution of major areas of cultivation of yams and African rice. The glaberrima rice has a semiendemic variation pattern.

Monocentric

Crops with a definable center of origin and wide dispersal without secondary centers of diversity. Examples: Arabica coffee and hevea rubber. Crops of this class are mostly new plantation or industrial crops. Ancient widespread crops usually develop secondary centers, but this takes time. .

Oligocentric

Crops with a definable center of origin, wide dispersal, and one or more secondary centers of diversity. Examples: the whole Near East complex of barley, emmer, flax, pea, lentil, oats, chickpea, *Brassica* spp., and so on; all have secondary centers in Ethiopia and some also have secondary centers in India and/or China.

Noncentric

Crops whose patterns of variation suggest domestication over a wide area. The suggestion may be misleading, of course, but centers are either not apparent or anomalous. Examples: sorghum, common bean, and field mustard.

Noncentric Crops

Aside from the subjective questions of how endemic is "endemic" and how widespread is "widespread," the categories are clear and self-evident except for the one called "noncentric." It seems to me that some crops simply do not have centers, and the concept of centers of either origin or diversity as universal phenomena can be called into question.

Sorghum is selected as an example of a noncentric crop; that is, neither a center of diversity nor a center of origin is evident from the distribution of variation alone. Vavilov had indicated that Ethiopia was a center of diversity–center of origin for sorghum, but he did not know sorghum well and did not understand African crops. To some extent, Ethiopia is a center of diversity for the durra race and is certainly the main center of diversity for the durra–bicolor race, but these are only 2 of 15 races according to the classification of Harlan and deWet (1972). All other cultivated races are rather poorly represented in Ethiopia.

The region from eastern Nigeria through Chad and western Sudan is a center of diversity for the caudatum, guinea–caudatum, and durra–caudatum races, but not for the races of Ethiopia. The region from western Nigeria to Senegal is the center of diversity for the guinea race, and while there are some durras in the drier zones, other races are poorly represented. The area from Tanzania to South Africa is the center for the kafir race and although guineas and guinea–kafirs are found, they are not especially variable. There is no area in Africa where the diversity of even several races is highly concentrated.

Snowden (1936) suggested that the several major races of sorghum had been ennobled separately from distinct wild races, but more recent DNA studies suggest that all five races derived from (*Sorghum bicolor* (L.) Moench subsp. *verticilliflorum* (Steud.) de Wet ex Wiersema & J. Dahlb) (Winchell et al., 2017). However, after the initial domestication of *S. bicolor*, the races may have been created following introgression of wild forms into *S. bicolor*. Multiple ennoblements or formation of introgression hybrid races are probably common, however, and must be dealt with by geographers and students of plant domestication.

The patterns of distribution for the races of *Sorghum* are remarkably consistent and clear-cut and presumably mean something with respect to the origin and evolution of sorghum. Harlan and Stemler (1976) attempted to reconstruct the history of sorghum domestication using this and other distributional information. The results are shown in Figures 7.3 and 7.4. While we show a center of origin labeled "Early Bicolor" in Figure 7.3, this region was not selected because of any clues given by variation patterns in modern cultivated sorghum. It was chosen because (a) archaeological evidence suggests African agriculture originated north of the equator, (b) the West African race of wild sorghum is a forest grass with an adaptation quite different from that of the crop, and (c) the region outlined includes the most

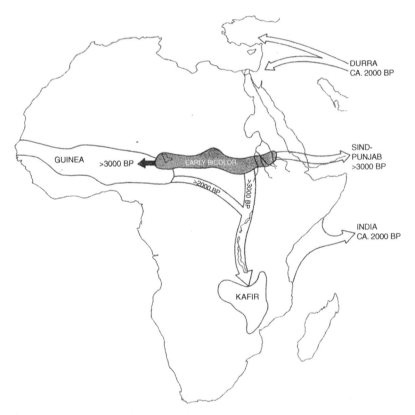

Figure 7.3 Suggested early movements of sorghum. Initial domestication in shaded area (from Harlan and Stemler, 1976).

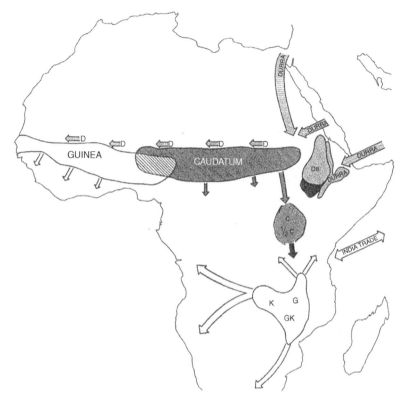

Figure 7.4 Suggested later movements of sorghum in Africa (from Harlan and Stemler, 1976). Abbreviations: C = Caudatum, D = Dura, G = Guinea, K = Kafir, D8 = Dura-Bicolor, GK = Guinea-Kafir.

massive stands of wild sorghum adapted to a savanna habitat north of the equator. These conclusions were confirmed from DNA evidence reported by Winchell et al. (2017) who showed that the origin of *S. bicolor* was in the central eastern Sudan region.

Today the most commonly cultivated race in the region is caudatum, which we consider to be a relatively new race. One could describe the pattern in terms of a series of secondary centers, for example, West Africa for guinea, southern Africa for kafir, Sudan–Chad–Uganda for caudatum, and Asia–Ethiopia for durra, but these are not centers of diversity for sorghum; they are centers for only certain races of sorghum. Some crops clearly have centers and others do not. Variation patterns must be analyzed in each crop separately before generalizations can be made.

The idea of a noncentric crop is not new and was well documented by a distinguished member of Vavilov's professional team. E. N. Sinskaya (1928) wrote a monograph on some of the cruciferous crops; she made the following observation about rapeseed (syn. colza):

> The geographical distribution of the forms of colza, as it may be pictured on the basis of data available at the present time, points in no way to the existence of a special center of diversity. To every region corresponds a definite ecotype. The introduction into cultivation has taken place, and is still taking place, in every region independently of any "center." The cultivated forms are identical with the local weeds; the local climatical ecotype being first distributed as weed, becomes afterward a cultivated crop.

Nothing since has been found to change that impression. Indeed, the introduction of the *Brassica* weed into cultivation, having taken place over most of Eurasia, appears to be once again in process in the Andean highlands of South America (Gade, 1972).

The common bean (*Phaseolus vulgaris* L.) has a pattern rather similar to the examples just given. In South America, wild races are found along the eastern slopes of the Andean mountain chain at mid-elevations from Argentina to Venezuela, about 5000 km. Brücher (1968) demonstrated that different races were domesticated at different places and at different times along this distribution. Gentry (1969), however, showed another area of bean domestication that centers in south and western Mexico; and Gepts et al. (1988) have convincing confirmation of at least two sources of bean germplasm from biochemical studies. Common bean and lima beans were thus domesticated at least twice, once in Mesoamerica and a second time in South America (Gaut, 2014; Motta-Aldana et al., 2010) whereas scarlet runner bean and tepary bean were only domesticated in Mesoamerica (Kaplan, 1965; Kaplan & Lynch, 1999). Altogether, the area of domestication stretched over 7,000 km with a series of local domestications.

Diffuse Origins

Crop origins can be diffuse in both space and time. Even if a crop enters the domestic fold in a new, but limited area, it may change radically as it is dispersed from its center of origination. As it spreads, it may receive infusions of germplasm from its wild relatives, and people in different regions may apply very different selection pressures. The most highly derived end products may be far removed geographically and morphologically from the wild progenitors from which they evolved.

Maize was domesticated from the Balsas teosintes in southern Mexico (Matsuoka et al., 2002). Cultivation of maize spread slowly in all directions from its center of origin (Bedoya et al., 2017; Hufford et al., 2012). At the time of European contact, it was being cultivated from southern Canada (about 50 °N) to southern Argentina (about 43 °S) and Chile and throughout the Caribbean Islands. Each region had its own characteristic array of races. Some rather large areas had only a few races, and these were relatively uniform; other, much smaller regions were characterized by many races, some of which were highly variable. These areas of diversity occurred in southern Mexico, Guatemala, parts of Colombia, and Peru. In fact, Peru is noted for extreme diversity in maize. The giant-seeded Cuzco, forms with "interlocking" cobs, cultivars with extremely long and flexible cobs, and so on, are among the unique races of the region. These innovations are not found in the center of origin, but far removed from it, and reflect the spread of large effective populations, the presence of gene flow from other teosintes (Ross-Ibarra et al., 2009; Warburton et al., 2011), and a very plastic genome full of redundant DNA sequences and transposable elements (Makarevitch et al., 2015a, 2015b). Despite arguments made over the years, there was only one domestication event of maize, and only one direct ancestor (teosinte).

All the available information indicates that barley was first domesticated in the Near East, and the present distribution of wild barley together with archaeological evidence point to a rather specific part of the region (see Chapter 8). It has become, perhaps, the most widely grown of major crops, being cultivated from above the Arctic Circle to southern Argentina and Chile as well as in tropical latitudes. The progenitor is the wild two-rowed *Hordeum vulgare* L. subsp. *spontaneum* (K. Koch) Thell., and the earliest barleys from archaeological sites are two-rowed. Two-rowed barleys are still common throughout the region and are grown in the drier zones under rainfed conditions. The irrigated barleys are usually six-rowed.

Variation is not notable in the center of origination, and most of the races of barley occur elsewhere. Each geographic and ecological region has its own set of cultivars with characteristic concentrations of particular genetic traits. The Ethiopian plateau is especially favorable for the development of leaf diseases, and the barleys have responded over the centuries by developing high frequencies of genes for resistance. Genes conditioning *irregular* (seed formed in some of the lateral spikelets) and *deficiens* (lateral spikelets suppressed) head types are also common in Ethiopian barleys. Barleys of Tibet and adjacent highlands show high frequencies of the gene for naked seeds and some are hooded. Chinese and Japanese barleys have their own peculiar characteristics.

However, Ethiopian barleys originated in Ethiopia, Tibetan barleys in Tibet, and Chinese barleys in China. All of them differ considerably from the primitive two-rowed barleys first cultivated in the Near East. In this case, we have a center of origin that can be located within reasonable limits, but it is obvious that most of the evolution of barley took place elsewhere. To say that barley, as we now know the crop, originated in the Near Eastern nuclear area is misleading, to say the least.

Wheat is an even more extreme example, for what originated in the center of origin was not wheat as we know it, but einkorn and emmer, crops that are now obsolete. There is good reason to believe that hexaploid bread wheat originated outside of the nuclear area where einkorn and emmer were first cultivated. Again, we have a crop dispersed over vast geographic areas and evolving new arrays of locally adapted cultivars as it spread. Wheat is relatively uniform over extensive parts of its range of distribution, but in specific local regions there are nodes of variation or centers of diversity.

Variation in a crop may be increased considerably if the crop is used for different purposes by different people. The common bean, for example, may be selected for green beans or for dry beans. The garden pea may be selected for green peas, dry peas, or edible pods. The mung bean is ground into flour in India and sprouted in China. *Corchorus olitorius* is a vegetable in Africa and a fiber in India. Roselle is a food in Africa and a fiber in India. Some people grow flax for fiber, some for edible seeds, and some for industrial oil. Hemp is grown for fiber, edible seeds, and narcotics. Most cereals have multiple uses and special races are developed accordingly. The variation patterns of crops are largely artifacts resulting from human activity; therefore, the larger the number of people who grow a crop and the greater their diversity, the more variable the crop is likely to be.

Microcenters

Even within centers of diversity, it is not unusual for a crop to be reasonably uniform over extensive areas and to show enormous diversity in very small regions. This is the microcenter phenomenon that I described many years ago (Harlan, 1951). Later the term was used in a very different sense by Zhukovsky (1968). Microcenters, as I originally observed them, are relatively small regions, 100–500 km across, in which may be packed an astonishing variation of one to several crops. I have observed microcenters for wild plants as well (Harlan, 1963a, 1963b), and they appear to be fairly common in the variation patterns of plant species.

Variability is of such an order that microcenters can be spotted easily in the field, and it contrasts sharply with the amount of variation in adjacent regions. The pattern has not been investigated as much as it deserves, but in the microcenters that I studied, the source of variability appeared to involve introgression between contrasting populations.

As of the mid-1900s, one could still detect microcenters scattered across the Near East: Turkish Thrace, Transcaucasia with adjacent parts of Turkey, parts of Iran, and Afghanistan. Some of them have been destroyed, and the rest are threatened by replacement with modern cultivars. Most of them were located outside of the Near Eastern nuclear area and the source of variation was attributed to current evolutionary activity rather than to any relationship to crop origins.

Landrace Populations

For people accustomed to modern agriculture, it seems necessary to describe landrace populations. It is only within the last century that landraces have been replaced by uniform, true-breeding cultivars or special hybrids of controlled parentage. Traditionally, field crops consisted of landrace populations rather than cultivars in the modern sense. Landraces are still grown, of course, wherever traditional agriculture is practiced.

Landrace populations are often highly variable in appearance, but they are each identifiable and usually have local names. A landrace has particular properties or characteristics. Some are considered early maturing and some late. Each has a reputation for adaptation to particular soil types according to the traditional peasant soil classification, for example: heavy or light, warm or cold, dry or wet, strong or weak. They may also be classified according to expected usage; among cereals, different landraces are used for flour, porridge, "bulgur" (food made from cracked parboiled groats of wheat), and malt to make beer, and so on. All components of the populations are adapted to local climatic conditions, cultural practices, and diseases and pests.

Genetic variation within a landrace may be considerable, but it is far from random. The various component genotypes have survived in a region for a long period of time or else they are offspring of lines that have undergone local selection for many generations. The genotypes are not only adapted to their environment, both natural and man-made, but they are also adapted to each other. A landrace population is an integrated unit and the components have adjusted to one another over the generations. Landraces are adapted to conditions of traditional agriculture, they are adapted to low soil fertility, low plant populations, and low yield. On the other hand, the genetic variability provides some built-in insurance against hazards. Devastating disease epidemics are unlikely because the populations contain such an array of resistance genes that no single race of pathogen can build up to epidemic proportions. Some genotypes would be affected each year, but not all of them.

Seedlings of landraces can emerge from a wide range of planting depths. Landraces usually produce a harvestable crop despite drought, standing water, insects, hail, or diseases. In traditional agriculture, high yields have never been necessary, but a crop failure means famine and death. Landraces may not yield much by modern standards, but they tend to be dependable.

The composition of landraces is frequently deliberately manipulated by cultivators. In Africa, the first step at sorghum harvest time is the selection of seed stocks for the next planting season. The farmer walks through his field and carefully chooses heads to be saved for seeding. Sometimes the heads are rather similar; sometimes a remarkable array of head and seed types is assembled. I have often asked the reasons for choosing a particular type. There is always a reason, but it may have little relation to the answers I received. When a farmer selects a variable range of material, the reason given is usually to the effect that a mixture of types is more nutritious than uniform strains. This could well be true. On one occasion, I noticed that a farmer had selected crook-necked types from his field. On inquiry, he replied that they are easier to hang from the tukel roof. [It is common to store the stock seed inside the house (tukel) above the hearth where the smoke from the kitchen fires provides some protection from weevils.]

Similar procedures are used by traditional farmers for maize. Ears are carefully chosen for stock seed and set aside. The reasons for selecting this or that are complex and deserving of serious anthropological and genetic study. Wellhausen et al. (1952) suggested that the very intense pigmentation of highland *elote* maize came about because of the usage of these races for roasting ears rather than because of some esthetic appreciation of colored seeds. Mexicans prefer floury endosperm in roasting ears, and it is easier to keep stocks pure for floury if the more intensely pigmented types are selected. In Peru, some deeply pigmented races are selected for the production of a red, nonalcoholic beverage (chica morada). Anderson (1954) noted that some farmers in Mexico sow an occasional seed of red-eared maize in their fields of white or yellow races "for luck." H. G. Wilkes (1977) observed Mexican peasant farmers actually building hybrid races by interplanting selected parental types. The practice might well be very ancient. The genetic consequences of these behavior patterns have not been seriously studied.

Whatever the reasons for the choice and whether the selections tend to be uniform or variable, all the components are from adapted materials. Landraces are built up and the components are selected, reassorted, recombined, and rearranged, but the local materials are constantly being adjusted to local conditions. The great variability of landraces makes them good sources for genes for modern plant breeding. They are not adapted to high fertility, high plant populations, or high production, but their dependability makes them useful in difficult or marginal situations and to introduce variation into the breeding population.

Implications for Plant Breeding

Analyses of variation patterns of crops are essential to understand the germplasm that went into their evolution and to make efficient use of the available variability in plant breeding. Typical variation patterns include:

1) Wild populations that are often highly variable, especially when they cover a considerable geographic range and/or ecological amplitude.
2) Landrace populations that are balanced, integrated mixtures of genotypes adapted to a region and to cultural practices in vogue.
3) Weed populations, frequently derived from genetic interaction between wild and cultivated races.
4) Microcenters in which enormous diversity is found in a restricted geographic area, usually due to genetic interaction between cultivated races and/or spontaneous races.
5) Secondary centers in which great variation has accumulated in certain special geographic regions, usually with considerable isolation from other regions for long periods of time.

Depending on the age and distribution of the crop, the variation may have an endemic, semiendemic, monocentric, oligocentric, or noncentric pattern.

Geographic patterns of variation help direct plant exploration and germplasm assembly for breeding programs. Collecting is most rewarding in centers of diversity and in microcenters, when they can be found. We now know that certain geographical areas may have concentrations of genes for multiple disease resistance. Other useful traits have been plotted, and we know enough about variation patterns in some crops so that reasonably systematic collections can be made. Not all of the useful genes are found in centers of diversity. Cultivars being grown near the climatic or ecological limits of a crop may have special attributes. Sorghum on the high plateaus of Ethiopia are especially cold tolerant; pearl millet in southern Chad, growing under higher rainfall than pearl millet usually receives, is particularly resistant to leaf diseases; barley from high elevations in the Himalayas is more resistant to frost damage at flowering time than other races, and so on.

Traditional farmers, over the millennia, have given us a priceless heritage of germplasm of major and minor crops. Simply by tilling the soil and cultivating plants they have developed a vast array of diversity in plants that are essential to our civilization. The heritage is in peril because of recent developments. The ancient patterns of variation are being obliterated. The first expression of concern I can find in the literature was expressed by my father, H. V. Harlan, and his coworker, M. L. Martini, in 1936 (p. 317):

> In the great laboratory of Asia, Europe, and Africa, unguided barley breeding has been going on for thousands of years. Types without number have arisen over an enormous area. The better ones have survived. Many of the surviving types are old. Spikes from Egyptian ruins can often be matched with those still growing in the basins along the Nile. The Egypt of the pyramids, however, is probably recent in the history of barley. . . . In the hinterlands of Asia there were probably barley fields when man was young. The progenies of these fields with all their surviving varieties constitute the world's priceless reservoir of germplasm. It has waited through long centuries. Unfortunately, from the breeder's standpoint, it is now being imperiled. . . . When new barleys replace those grown by farmers of Ethiopia or Tibet, the world will have lost something irreplaceable. When that day comes our collections, constituting as they do but a small fraction of the world's barleys will assume an importance now hard to visualize.

The threat was not generally evident until after World War II. A small program was started in Mexico through an agreement between the Rockefeller Foundation and the Mexican Department of Agriculture. It was a modest start in international cooperation in agriculture research, but the consequences were far reaching and unexpected. The program involved wheat, maize, beans, forage crops, and so on, but the wheat program under the direction of Norman Borlaug had the greatest initial impact. Most improved cultivars usually yield only a few percent more than the ones they replace. Borlaug's wheats yielded as much as four times the traditional landraces, with suitable inputs. The result was an extraordinary increase in wheat production in Mexico and a worldwide demand for high yielding varieties. The new varieties soon began to sweep the world, especially in developing countries where local wheat diversity was greatest. Concern began to be expressed for the germplasm being displaced.

The success of the initial program generated interest in expansion to other countries and other crops. Much more money had to be generated and international institutes were established (see Table 7.1). The International

Table 7.1 The international centers with major crop genetic resource holdings.

Abbreviation & year established	Institute	Major collections
IRRI (1960)	International Rice Research Institute, Los Baños, Philippines	Rice
CIMMYT (1967)	Centro Internacional de Mejoramiento de Maíz y Trigo, El Batán, México, D.F. México	Maize, wheat
IITA (1968)	International Institute for Tropical Agriculture, Ibadan, Nigeria	Bananna, cassava, cowpea, yam, soybean, bambara groundnut, maize
CIAT (1969)	Centro Internacional de Agricultura Tropical, Cali, Colombia	Beans, tropical forages, cassava
AfricaRice (1971)	African Rice Center, Abidjan, Ivory Coast	African rice
CIP (1972)	Centro Internacional de las Papas, Lima, Peru	Patoto, sweet potato, Andean root & tuber crops
ICRISAT (1972)	International Crops Research Center for the Semi-Arid Tropics, Hyderabad, India	Peanut, chickpea, millets, sorghum, pigeon pea
Bioversity (1974)	Bioversity, International, Maccarese, Italy	*Musa* species
ICARDA (1976)[*]	International Center for Agricultural Research in Dry Areas	
	Morocco	Cultivated barley, chickpea, grasspea, lentil, wheat
	Lebanon	Wild relatives of cereals & legumes
ICRAF (1978)	World Agroforestry, Nairobi, Kenya	Fruits, tree species
Svalbard (2008)	The Svalbard Global Seed Vault, Longyearbyen, Norway	Long term storage

[*]The ICARDA genebank was destroyed in the Syrian Civil War which started in 2011; collections were moved to Morocco and Lebanon.

Rice Research Institute was the first to be established and was located in the Philippines to do with rice what Borlaug had done with wheat. Afterward, nine additional centers were created.

IRRI succeeded in creating highly productive rice and soon these culti-vars were replacing landraces in the very centers of rice diversity. The level of concern about germplasm began to rise. Other crops were having similar successes. International concern finally demanded some action.

The Food and Agriculture Organization of the United Nations (FAO) con-vened a major meeting in Rome in 1967, to discuss the problem and recom-mend action. There were more meetings and more planning and more discussions. Papers were written and books were published, but finally an International Board for Plant Genetic Resources (IBPGR; now called Bioversity International) was established in 1974 under the Consultative Group on International Agricultural Research (CGIAR) and funded through the same sources as the international institutes.

Collectors were sent out to sample germplasm of various crops. Curators of collections were appointed and gene banks were established. In theory, the effort was to save the genes before they disappeared. In fact, a lot of collections were made. How well we saved the genes we will never know. We now have some very large holdings, but little use of them is being made by plant breeders because evaluation has been slow. Phenotypic characterization also has been slow and underfunded, and while new DNA sequencing technologies have speeded genetic characterization in many species, a lot of work must still be done before these collections are put into usable condition. Did we save the genes? Surely not all of them, but we are doing our best.

The ancient patterns of diversity have been largely destroyed. Our herit-age of crop diversity is preserved, for better or worse, in cold storage. To be sure, there are still traditional farmers growing traditional landraces. There have been repeated proposals that encourage farmers to continue preserv-ing germplasm *in situ*. The argument is that landraces have evolved over millennia in consort with their pathogens and pests and biological balances have been achieved. The coevolution would continue into the future if only we preserve the populations *in situ*. Collecting the landraces and putting them in cold storage will freeze the evolution as well as the seed. True enough, but how to do it? Parks and preserves have been established to con-serve wild species, but so far little has been done to conserve landraces of important domesticates *in situ*.

The concept of *in situ* conservation has genetic merit, but the sociological problems are complex. A farmer has the right to grow high-yield cultivars if he chooses. On the other hand, some farmers choose to retain the old lan-draces. The spread of modern cultivars has been slowing and we have prob-ably fairly well saturated the most productive regions. Research emphasis in the international institutes has been shifting toward marginal areas and the less-productive environments. Rice breeders are devoting attention to cold-water tolerance, salt-water tolerance, drought tolerance for upland rice, and deep-water tolerance. Wheat breeders are working on tolerance to toxic soils, heat and cold tolerance, and so on. Improvement in marginal areas is much more difficult than in the most productive regions. Consequently, traditional farmers in the marginal regions are more likely to maintain their traditional landraces. These farmers will probably provide living, dynamic, evolving gene banks well into the future. How we might encourage others to do the same remains to be worked out.

Recent activities have resulted in some huge collections, far more than plant breeders can handle in the absence of information about the accessions. There is serious concern about the management of these holdings. Who and how will they be rejuvenated when viability begins to wane? The U.S. National Seed Storage Laboratory (NSSL) has no facilities and no mandate for regeneration at all. When viability declines, all it can do is return the seed to the source. But the source may have retired, died, gone out of business, or have no resources either. However, in the 1990s, many of the NSSL holdings were put into cryopreservation, where they are expected to maintain viability for many decades to come. In addition to the NSSL, the USDA established 17 regional sites that employ germplasm curators to manage 595,451 (2019) accessions of crops and related wild species. These sites are working collections where new germplasm is deposited, stored as seeds or plants, regenerated, and distributed. Holdings in developing world countries may be even more vulnerable. Gene banks and number of accessions proliferated, but money for maintenance is often lacking. We have moved our precious heritage into gene banks, but how many banks will fail? In 2004, the Crop Trust was founded to help address some of these concerns. It is funded by an endowment, which allows funding for the international centers and other gene banks, to maintain worldwide collections in perpetuity. The Svalbard Seed Vault was created in 2008 and holds backups of many international and national collections, but there remains a significant amount of germplasm to be stored there. As of 2020, the Vault holds over 980,000 samples, which come from almost every country in the world.

On the other hand, the international institutes are, in fact, international in scope and have served a very important function in both assembly and distribution of genetic resources. They vary a good deal, however, in competence and philosophy for handling germplasm. In addition to the international institutes listed in Table 7.1, a sample of the major national gene banks sponsored by individual countries is listed in Table 7.2.

Table 7.2 A sample of the major national gene banks in the international system.

VIR	Vavilov Institute of Plant Industry, St. Petersburg, Russia
USDA	US Department of Agriculture, several locations, USA
NSSL	National Seed Storage Laboratory, Ft. Collins, CO, USA
IBBR-CRN	Institute of Biosciences and Bioresources, Bari, Italy
NordGen	Nordic Gene Bank, Lund, Sweden
IPK	Genebank, Gatersleben, Germany
PGRC	Plant Genetic Resources, Ottawa, Ottawa, Canada
NARO	National Agriculture and Food Research Organization Genebank, Tsukuba, Japan
ICGR	Institute of Crop Germplasm Resources, Chinese Academy of Agricultural Science, Beijing, China
ICAR	Indian Council for Agriculture Research, National Bureau of Plant Genetic Resources, New Delhi, India
Korea Genebank	National Agrobiodiversity Center, National Institute of Agricultural Sciences, Jeonju, South Korea
EMBRAPA	The Brazilian Agricultural Research Corporation, Rio de Janeiro, Brazil

The collections held in the national and international gene banks are a wonderful resource, but prone to organizational problems. First of all, there is a great deal of redundancy in the collections. There are many outright duplicates and there is excessive collection of common materials with little variation. This is likely due in large part because collectors followed the same roads to obtain plant materials and more remote areas were extremely difficult to sample. We have over-sampled some material and failed to collect some of the rarer and less accessible material. The monster collections would be much more usable if the duplicates and over-collected types were reduced in size. Two methods to deal with the problem are being implemented. As genotyping becomes cheaper, duplicates and over-representatives can be identified. Actual duplicates may be bulked into one entry, and over-representation can be reduced by establishing *core collections*. By carefully choosing accessions that sample diversity geographically, taxonomically, and genetically, a collection of 20,000 could easily be reduced to maintain nearly all of the diversity of the larger sample. Theories about how best to construct a core collection have been developed (Hodgkin et al., 1995; Upadhyaya et al., 2007; van Hintum et al., 2000; Wang et al., 2014). Core collections generally consist of 10% of the base collection and mini-core collections (1% of the base collection) and many have been developed for cereals, legumes, vegetables, and other crop groups. A plant breeder might make use of a 2000-item collection, but has neither the time nor resources to tackle 20,000.

To be sure, the rare genes (alleles) could easily escape a core sample, and there are times when rare genes are critically important. They seldom have anything to do with yield or adaptation, but are usually concerned with disease or pest resistance, or with chemical nulls (e.g., T. Hymowitz's work with soybean). To find the rare genes, the whole collection may need to be screened and should be kept available for the purpose. If the breeder screens the core collection first and does not find the gene, then he knows it is rare. Even so, it is not profitable to plow through the overcollected material. The rare gene is likely to be in less common material or at times, only found in related wild species.

One way to keep the evolution going is to make up composite crosses and grow them under different environments (Chapter 6). It is a somewhat different evolution than would take place in traditional agriculture, but if traditional agriculture is going to disappear, what can we lose by trying? It is a practical method for conserving genes, preserving diversity, and developing adaptive linkages and gene associations.

Much nonsense has been written about the "necessity" of preserving landrace populations in their original form. It has often been recommended that the population be taken back to their sources for increase and rejuvenation. In this way the original linkages would be preserved. It is a popular armchair theory, but in practice will never be done. If a plant breeder uses a traditional landrace in his program, the first thing he will do is to try by whatever means possible to break up the linkages as quickly as possible. The linkage of deleterious genes is the bane of the plant breeder and one reason he is reluctant to work outside his elite material. The plant breeder wants the genes, not the linkages.

Much more could be said about genetic resource management, but that is beyond the scope of this book. Literature on the subject is extensive. Some older reference sources are: Bakhteev (1960), Brown et al. (1989, 1990), Burgess (1971), Frankel (1973), Frankel and Bennett (1970), Frankel and Soulé (1981), Holden and Williams (1984), Kloppenburg (1988), National Academy of Sciences (1972), National Research Council (US) Committee on Managing Global Genetic Resources: Agricultural Imperatives (1991), and Plucknett et al. (1987). A few of the many more recent references include Fu (2017), Namkoong et al. (2005), Singh (2007), and van Treuren and van Hintum (2014).

Conclusions

The Vavilovian concepts of "centers of origin" were too simplistic. It is necessary to examine each crop separately. When this is done, it becomes obvious that there is a variety of geographical patterns depending on the history and distribution of each crop and its wild progenitor(s). Some had centers of origin; some did not. Some were domesticated at least several times, others only once. Some spread early and developed secondary centers; some spread recently and can be traced to their origins by historical data. Each crop was shaped and molded throughout its history by human activities, by uses, preferences, cultural practices, and by continuous adjustments to the environment provided, including climate, soils, agronomic management, diseases and pests. Space, time, and variation are all part of the geography of crop plants and fundamental to the collection and preservation of crop genetic resources.

References

Anderson, E. (1954). *Plants, man and life*. London: A. Melrose.

Bakhteev, F. K. (1960). *The world resources of the useful plants* [In Russian]. Moscow, Russia: Publ. Acad. Sci. SSSR.

Bedoya, C. A., Dreisigacker, S., Hearne, S., Franco, J., Mir, C., Prasanna, B. M., . . . Warburton, M. L. (2017). Genetic diversity and population structure of native maize populations in Latin America and the Caribbean. *PLoS One, 12*(4), E0173488. https://doi.org/10.1371/journal.pone.0173488

Brown, A. H. D., Clegg, M. T., Kahler, A. L., & Weir, B. S. (1990). *Plant population genetics, breeding, and genetic resources*. Sunderland, MA: Sinauer Associates Inc.

Brown, A. H. D., Frankel, O. H., Marshall, D. R., & Williams, J. T. (1989). *The use of plant genetic resources*. Cambridge, UK: Cambridge University Press.

Brücher, H. (1968). Die evolution der gartenbohne *Phaseolus vulgaris* L. aus der sudamerikanischen Wildbohne *Ph. aborigineus*. Buck. *Angewandte Botanik, 42*, 119–128.

Burgess, S. (1971). *The national program for conservation of crop germ plasm*. Athens, GA: University of Georgia Printing Department.

Callen, E. O. (1967). The first New World cereal. *American Antiquity, 32*, 535–538. https://doi.org/10.2307/2694082

Darlington, C. D., & Janaki Ammal, E. K. (1945). *Chromosome atlas of cultivated plants*. London: G. Allen and Unwin.

Frankel, O. H. (1973). *Survey of crop genetic resources in their centres of diversity: First report*. Rome, Italy: IBP/FAO.

Frankel, O. H., & Bennett, E. (1970). *Genetic resources in plants—their exploration and conservation.* Philadelphia, PA: F.A. Davis Co.

Frankel, O. H., & Soulé, M. E. (1981). *Conservation and evolution.* Cambridge, UK: Cambridge University Press.

Fu, Y. (2017). The vulnerability of plant genetic resources conserved ex situ. *Crop Science, 57,* 2314–2328. https://doi.org/10.2135/cropsci2017.01.0014

Gade, D. W. (1972). Setting the stage for domestication: *Brassica* weeds in Andean peasant ecology. *Proceedings of the Association of American Geographers, 4,* 38–40.

Gaut, B. S. (2014). The complex domestication history of the common bean. *Nature Genetics, 46,* 663–664. https://doi.org/10.1038/ng.3017

Gentry, H. S. (1969). Origin of the common bean *Phaseolus vulgaris. Economic Botany, 23,* 55–69. https://doi.org/10.1007/BF02862972

Gepts, P., Kmiecik, K., Pereira, A., & Bliss, F. A. (1988). Dissemination pathways of common bean (*Phaseolus vulgaris,* Fabaceae) deduced from phaseolin electrophoretic variability. I. The Americas. *Economic Botany, 42,* 73–85. https://doi.org/10.1007/BF02859036

Harlan, H. V., & Martini, M. L. (1936). *Problems and results of barley breeding. USDA yearbook of agriculture* (pp. 303–346). Washington, DC: U.S. Government Printing Office.

Harlan, J. R. (1951). Anatomy of gene centers. *The American Naturalist, 85,* 97–103. https://doi.org/10.1086/281657

Harlan, J. R. (1963a). Two kinds of gene centers in Bothriochloininae. *The American Naturalist, 97,* 91–98. https://doi.org/10.1086/282259

Harlan, J. R. (1963b). Natural introgression between *Bothriochloa ischaemum* and *B. intermedia* in West Pakistan. *International Journal of Plant Sciences (Chicago, IL, United States), 124,* 294–300. https://doi.org/10.1086/336209

Harlan, J. R., & deWet, J. M. J. (1972). A simplified classification of cultivated sorghum. *Crop Science, 12,* 172–176. https://doi.org/10.2135/cropsci197 2.0011183X001200020005x

Harlan, J. R., & Stemler, A. B. L. (1976). The races of sorghum in Africa. In J. R. Harlan, J. M. J. deWet, & A. B. L. Stemler (Eds.), *The origins of African plant domestication.* The Hague, The Netherlands: Mouton. https://doi. org/10.1515/9783110806373.465

Hodgkin, T., Brown, A. H. D., van Hintum, T. J. L., & Morales, E. A. V. (Eds.) (1995). *Core collections of plant genetic resources.* Rome, Italy: International Plant Genetic Resources Institute (IPGRI).

Holden, J. H. W., & Williams, J. T. (1984). *Crop genetic resources: Conservation and evaluation.* London: Allen and Unwin.

Hufford, M. B., Martínez-Meyer, E., Gaut, B. S., Eguiarte, L. E., & Tenaillon, M. I. (2012). Inferences from the historical distribution of wild and domesticated maize provide ecological and evolutionary insight. *PLoS One*, *7*(11), E47659. https://doi.org/10.1371/journal.pone.0047659

Kaplan, L. (1965). Archeology and domestication in American *Phaseolus* (beans). *Economic Botany*, *19*, 358–368. https://doi.org/10.1007/BF02904806

Kaplan, L., & Lynch, T. F. (1999). *Phaseolus* (Fabaceae) in archaeology: AMS radiocarbon dates and their significance for pre-Columbian agriculture. *Economic Botany*, *53*, 261–272. https://doi.org/10.1007/BF02866636

Kloppenburg, J. R. (1988). *Seeds and sovereignty: The use and control of plant genetic resources*. Durham, NC: Duke University Press.

León, J. (1964). *Plantas alimenticias andinas. Boletin Tecnko No. 6*. Lima, Peru: Instituto Interamericano de Ciencias Agrícolas Zona Andina.

Makarevitch, I., Waters, A. J., West, P. T., Stitzer, M., Hirsch, C. N., Ross-Ibarra, J. R., & Springer, N. M. (2015a). Transposable elements contribute to activation of maize genes in response to abiotic stress. *PLoS Genetics*, *11*(1), E1004915. https://doi.org/10.1371/journal.pgen.1004915

Makarevitch, I., Waters, A. J., West, P. T., Stitzer, M., Hirsch, C. N., Ross-Ibarra, J. R., & Springer, N. M. (2015b). Correction: Transposable elements contribute to activation of maize genes in response to abiotic stress. *PLoS Genetics*, *11*(10), e1005566. https://doi.org/10.1371/journal.pgen.1005566

Matsuoka, Y., Vigouroux, Y., Goodman, M. M., Sanchez, J. G., Buckler, E., & Doebley, J. (2002). A single domestication for maize shown by multilocus microsatellite genotyping. *Proceedings of the National Academy of Sciences of the United States of America*, *99*, 6080–6084. https://doi.org/10.1073/pnas.052125199

Motta-Aldana, J. R., Serrano-Serrano, M., Hernández-Torres, J., Castillo-Villamizar, G., Debouck, D., & Chacón, M. (2010). Multiple origins of lima bean landraces in the Americas: Evidence from chloroplast and nuclear DNA polymorphisms. *Crop Science*, *50*, 1773–1787. https://doi.org/10.2135/cropsci2009.12.0706

Namkoong, G., Lewontin, R. C., & Yanchuk, A. D. (2005). Plant genetic resource management: The next investments in quantitative and qualitative genetics. *Genetic Resources and Crop Evolution*, *51*, 853–862. https://doi.org/10.1007/s10722-005-0776-0

National Academy of Sciences (1972). *Genetic vulnerability of major crops*. Washington, DC: National Academy of Sciences.

National Research Council (US) Committee on Managing Global Genetic Resources. Agricultural Imperatives (1991). *Managing global genetic resources: The U.S. National Plant Germplasm System*. Washington, DC: National Academies Press.

Plucknett, D. L., Smith, N. J. H., Williams, J. T., & Anishetty, N. M. (1987). *Gene banks and the world's food*. Princeton, NJ: Princeton University Press. https://doi.org/10.1515/9781400858118

Portéres, R. (1956). Taxonomie agrobotanique des riz cultivés *O. sativa* Linné et *O. glaberrima* Steudei. *Journal d'Agriculture Tropicale et de Botanique Appliquee, 3*(7–8), 341–856.

Ross-Ibarra, J., Tenaillon, M., & Gaut, B. S. (2009). Historical divergence and gene flow in the genus *Zea. Genetics, 181*(4), 1399–1413.

Singh, A. K. (2007). Conventional conservation of agricultural and horticultural crops diversity. In A. K. Singh, K. Srinivasan, S. Saxena, & B. S. Dhillon (Eds.), *Hundred years of plant genetic resources management in India* (pp. 191–210). New Delhi, India: ICAR.

Sinskaya, E. N. (1928). The oleiferous plants and root crops of the family Cruciferae. *Trudy po Prikladnoi Botanike, Genetike i Selektsii, 19*, 1–648.

Snowden, J. D. (1936). *The cultivated races of sorghum*. London: Adlard and Son.

Upadhyaya, H.D., Gwoda, C.L.L., Pundir, R.P.S., & Ntare, B.R. (2007). Use of core and mini core collections in preservation and utilization of genetic resources in crop improvement. In Intenational Crops Research Institute for the Semi-Arid Tropic (Ed.), *Plant genetic resources and food security in West and Central Africa. Regional Conference*, 26–30 April 2004. Ibadan, Nigeria. Patancheru, India: ICRISAT. http://oar.icrisat.org/4119/acc

van Hintum, T. J. L., Brown, A. H. D., Spillane, C., & Hodgkin, T. (2000). *Core collections of plant genetic resources. IPGRI Technical Bulletin No. 3*. Rome, Italy: International Plant Genetic Resources Institute.

van Treuren, R., & van Hintum, T. J. L. (2014). Next-generation genebanking: Plant genetic resources management and utilization in the sequencing era. *Plant Genetic Resources, 12*, 298–307. https://doi.org/10.1017/S1479262114000082

Vavilov, N. I. (1926). *Studies on the origin of cultivated plants*. Leningrad, Russia: Inst. Appl. Bot. Plant Breed.

Vavilov, N. I. (1951). Phytogeographic basis of plant breeding—the origin, variation, immunity and breeding of cultivated plants. *Chronica Botanica, 13*, 1–366.

Wang, J., Guan, Y., Wang, Y., Zhu, L., Wang, Q., Hu, Q., & Hu, J. (2014). A strategy for finding the optimal scale of plant core collection based on Monte Carlo simulation. *Scientific World Journal*, 503473. https://doi.org/10.1155/2014/503473

Warburton, M. L., Wilkes, G., Taba, S., Charcosset, A., Mir, C., Dumas, F., . . . Franco, J. (2011). Gene flow among different teosinte taxa and into the domesticated maize gene pool. *Genetic Resources and Crop Evolution, 58,* 1243–1261. https://doi.org/10.1007/s10722-010-9658-1

Wellhausen, E. J., Roberts, L. M., Hernandez, E., & Mangelsdorf, P. C. (1952). *Races of maize in Mexico.* Cambridge, MA: Bussey Institute of Harvard University.

Wilkes, H. G. (1977). Hybridization of maize and teosinte, in Mexico and Guatemala and the improvement of maize. *Economic Botany, 31,* 254–293. https://doi.org/10.1007/BF02866877

Winchell, F., Stevens, C. J., Murphy, C., Champion, L., & Fuller, D. Q. (2017). Evidence for sorghum domestication in fourth millennium BC Eastern Sudan: Spikelet morphology from ceramic impressions of the Butana group. *Current Anthropology, 58*(5), 673–683.

Zhukovsky, P. M. (1968). New centres of the origin and new gene centres of cultivated plants including specifically endemic micro-centres of species closely allied to cultivated species. *Botanicheskii Zhurnal (Sankt-Peterburg, Russian Federation), 53,* 430–460.

8

The Near East

Why has Ea caused man, the unclean
To perceive the things of Heaven and Earth
A mind cunning has he bestowed on him
And created him into fame.
What shall we do for him?
Bread of life get for him; let him eat.

Akkadian Epic, third millennium BC
(Langdon & MacCulloch, 1931)

I kept alive Hefat and Hormer ... at a time
when . . . everyone was dying of hunger on
this sandbank of hell . . . All of Upper Egypt

Source: A logo of the Crop Evolution Laboratory, University of Illinois, Urbana, IL.

Harlan's Crops and Man: People, Plants and Their Domestication, Third Edition.
H. Thomas Stalker, Marilyn L. Warburton, and Jack R. Harlan.
© 2021 American Society of Agronomy, Inc. and Crop Science Society of America, Inc.
Published 2021 by John Wiley & Sons, Inc.
doi:10.2135/harlancrops

*was dying of hunger to such a degree that
everyone had come to eating his children,
but I managed that no one died of hunger
in this nome.*

Inscription on tomb of Ankhtifi, a nomarch
of Hierakonopolis and Edfu, approximately 4000 BP
(Bell, 1971)

Introduction

A considerable research effort has been concerned with an attempt to understand the shift from hunting–gathering economies to a food-producing economy in the Near East. Archaeologists, anthropologists, prehistorians, zoologists, botanists, geneticists, palynologists, geologists, and others have collaborated in team efforts to assemble evidence bearing on the problem. The research has resulted in a generalized framework of evidence that establishes a sequence of events and developments and the geographic regions in which they took place. As more and more information has accumulated, however, it is becoming increasingly clear that the process of developing an effective food-producing system was immensely complex and involved. The initial steps might have easily grown out of intensive gathering economies, but the end results were a complete revolution in food-procurement systems. A completely new ecological adaptation is not likely to be easily achieved.

The focus of the problem can be stated simply by beginning with the end products we know the most about. High civilizations did emerge in the Near East in Mesopotamia and in Egypt. These were not only the first of the high civilizations we know about, but because they provided the roots of Western civilization they have always been intensively studied and other civilizations have been measured, at least by Western man, according to the development of various elements found in the Near Eastern civilizations. The emergence of towns and cities, monumental buildings, professional classes, stratification of economic and political power, centralized government, priestly castes, standing armies, writing, and metallurgy are components of the criteria used to compare the Near Eastern civilizations with other civilizations.

During investigation of the emergence of Near Eastern civilizations, it became obvious that they were based on agriculture. City dwellers are consumers, not food producers, and agriculture is a prerequisite for any high civilization. To understand the origins of these civilizations then, one must understand how food production came into practice in the area.

Furthermore, an economic analysis of the historical period as well as earlier prehistoric communities showed that the bulk of the food produced and consumed came from the four domesticates: wheat, barley, sheep, and goats. There were other domesticated plants and animals, of course, but a high proportion of the caloric intake came from these four and the story of Near Eastern agriculture is largely their story.

Where traditional subsistence agriculture survives today, this situation still holds true. Cattle are used for work and milk; meat is a tertiary consideration and is more likely to be sold to city dwellers than consumed in the village. Domestic pigs are raised but are never much competition to sheep and goats as a source of food. Locally, dates, olives, and leguminous grains are important, but almost never replace the cereals as the staff of life.

In domesticating plants like wheat and barley, one presumably begins by harvesting the wild progenitor. Figures 8.1–8.3 show the distribution of known sites for wild barley, wild einkorn, and wild emmer, respectively. Stands of wild cereals can be extremely abundant today, and all three cereals may occur on extensive areas in patches as thick as a stand of cultivated grain. The wild emmer is abundant today in the Palestine area but also is found in Turkey, Iraq, Iran, and the former Soviet Union (currently, Russia, Turkmenistan, Uzbekistan, and Kazakhstan) in only thin, scattered stands. Experiments with harvesting the wild cereals have shown that it can easily be done with a flint-bladed sickle and that the effort would be rewarding in terms of food obtained per hour spent in harvesting (Harlan, 1967). Extensive natural stands of wild cereals would surely have been an attractive source of food for hunting–gathering people.

The archaeological problem, then, is to find out when and where the wild cereals were being harvested and when and where the first evidences of cultivated forms appear. Cereals are most likely to be found in archaeological sites in the form of carbonized grains or impressions in lumps of clay, mud walls, or adobe brick. Impressions and scraps of epidermis are not uncommon in pottery, but since the primary cereals were domesticated in pre-pottery times, this helps only in understanding later evolutionary events. The most useful specimens are usually the carbonized grains, although the changes in size and shape due to carbonization often cause problems in precise identification.

The most useful criterion in distinguishing wild from domesticated cereals is the articulation of the spikes or spikelets. Wild barley has a fragile rachis that shatters at maturity, whereas cultivated barley has a tough rachis that may remain more or less intact even after the seeds are threshed off the chaff. The wild wheats also have fragile rachises, but cultivated emmer and einkorn have rachises that are only somewhat tougher than the wild ones.

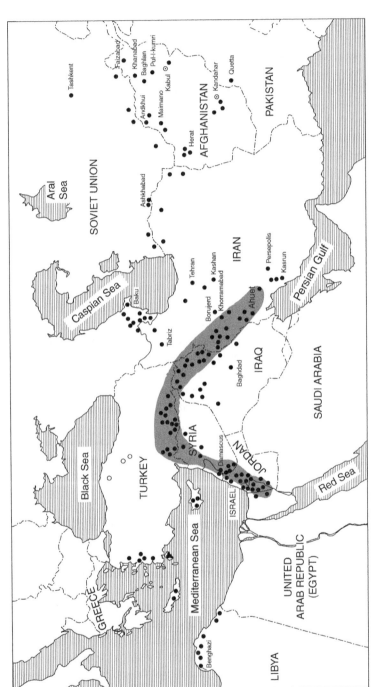

Figure 8.1 Distribution of known and reasonably certain sites of wild barley. Massive stands in fairly primary habitats may occur within a shaded area. Elsewhere, wild barley may be abundant, but confined to highly disturbed habitats. *Source:* Harlan and Zohary (1966). Reprinted with permission from AAAS. Isolated colonies have now been found in Tibet and Morocco.

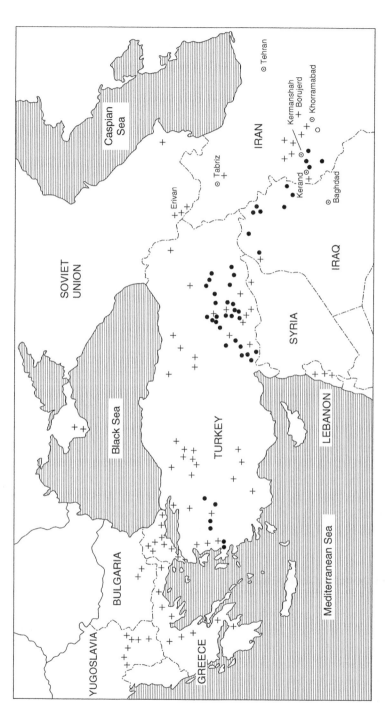

Figure 8.2 Distribution of known and reasonably certain sites of wild einkorn wheat—solid circles, fairly primary habitats; crosses, definitely segetal habitats. *Source:* Harlan and Zohery (1966). Reprinted with permission from AAAS.

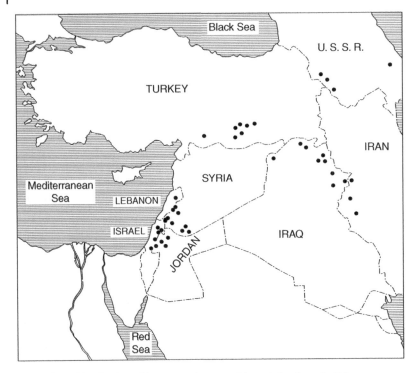

Figure 8.3 Distribution of known and reasonably certain sites of wild emmer. *Source:* Harlan and Zohary (1966). Reprinted with permission from AAAS.

The spike will remain intact until harvested, but on threshing breaks up at the joints just as in the wild forms. Additional processing is required to remove the seeds from the enveloping glumes. In the free-threshing wheats that evolved later, the rachis is tough, the glumes are deciduous, and the seed falls free when threshed. This character is a certain indication of domestication, as is the six-rowed character in barley.

Archaeological Prelude

The archaeological record of man in the Near East is respectably long, although not as long as that in Africa. Acheulean man (*Homo erectus Dubois*) apparently wandered over the whole region and left a scattering of his characteristic "hand axes." Later, Neanderthal man occupied many of the caves in the hilly or mountainous areas and left tools and artifacts of the Mousterian tradition. In the Near East, Neanderthal man bowed out about 35,000 years ago and there was a break in the evidence of human activity until the caves were reoccupied by anatomically modern man (*Homo sapiens*) just over 20,000 years ago. The artifacts he left evolved into Epipaleolithic traditions with striking regional differentiation. The tools and weapons became smaller, more efficient, and there was a trend toward compound pieces, that is, more than one blade mounted on a haft to form a sickle or several very small, razor sharp blades on a shaft to make a spear. Local differentiation increased with time.

The time of interest for the evolution of agriculture begins at the suture between the Pleistocene and the Holocene. For the Near East, this may be taken place about 12,000–11,000 years ago, somewhat earlier than for North America. The palynological information at that time range lacks data from most of Syria, all of Iraq, Saudi Arabia, and Jordan, but what we do have indicates a floral distribution strikingly different from the recent past (Van Zeist & Dottema, 1982).

At that time, the oak woodland was confined to a narrow strip along the Mediterranean in the Levant. There was a forest steppe or steppe with scattered trees over the western third of the Anatolian Plateau and most of Greece and a belt of true deciduous forest along the Black Sea coast of Turkey and the southern foreshore of the Caspian. The oak forest of the Zagros and Taurus as we know it today was missing. This formation is now found mostly within the shaded area (Figure 8.1). In the early Holocene, the area was covered by an Artemisia steppe vegetation. Wild emmer, einkorn, and barley are all found in the oak woodlands today, but not confined to them. Einkorn may range above the oak belt, and wild barley extends well below it into the deserts. The wild emmer, however, is rather closely associated with the oak belt, and all the early farming village sites have yielded remains of emmer. The movement of the oak belt into the hill country of the Zagros and Taurus was critical in setting the scene for the first plant domestications of the region. At least some of the oaks were in place by 10,000 years ago, and this marks the approximate beginning of the Pre-pottery Neolithic A (PPNA) era and the first evidence of domesticated plants.

A study of Figures 8.1–8.3 shows that all three wild cereals have discontinuous distributions. The implication is that the distributions were once continuous and later became discontinuous as a result of a change in climate. If, as a result of late Ice Age influence, the life zones had been displaced downslope and toward the south, all three species may have had continuous ranges. The outpost of wild einkorn in Lebanon might have been connected with the major range of the species and this may well account for the wild einkorn found at Mureybit and Abu Hureyra where wild einkorn does not occur today.

A Note About Dating Archaeological Sites

The most common method of dating archaeological sites is by ^{14}C. This isotope is formed by the interaction of neutrons from cosmic rays and ^{14}N. It is taken up by plants in photosynthesis and passed on to animals that eat the plants and becomes incorporated into bone, horns, shell, or any organic matter that survives archaeological time. Plant materials are generally more reliable than animal remains for dating. Carbon-14 decays to ^{14}N slowly, with a half-life of 5730 ± 40 years. It is assumed that the rate of decay is constant over time and not affected by either environment or time. Carbon incorporated into an organism 5700 years ago, then, should have about one-half the background percentage of ^{14}C remaining. The isotope can be detected in very small amounts, but with increasing age the tests become less and less accurate. By about 50,000 years ago, the dates are sufficiently inaccurate that other methods are preferred.

While the rate of decay is constant, as far as we know, the flux of cosmic rays apparently has varied over time causing variation in ^{14}C concentration. This was discovered by testing samples of wood of various ages from old trees. Tree ring dating is far more accurate than ^{14}C dating and where good material is available, approaches absolute dates with very little error. The oldest trees known are bristle cone pines of the US southwest. Carbon-14 dates can be corrected over certain time ranges by use of curves generated from samples taken from various parts of the trunk. Unfortunately, the curves are sinuous and are neither a straight line nor a curve with a constant function. If one really knew the date within narrow limits, a correction can be made, but if one has only an approximate date within broad limits, the correction could be worse than the uncorrected date. Carbon-14 dates are often sufficiently inaccurate that corrections are probably not warranted and are presented as uncorrected. The uncorrected dates are too young, and the greater the time range, the greater the discrepancy. Most people make a mental note of this, but use the dates as reported because they are extremely valuable on a comparative basis. A ^{14}C date of 5000 years in China should be equivalent to a ^{14}C date of 5000 years in Mexico or Europe.

Dates are often published with standard deviations, that is, plus or minus so many years. These refer to the accuracy of the laboratory test and do not mean much in terms of a range of absolute dates. They should not be ignored; they do tell something about the quality of the material submitted and the reliability of the results. In the early days of ^{14}C dating, different laboratories often came up with significantly different results from identical samples. The techniques are much improved today, and there are many competent laboratories giving reliable results on the samples submitted.

The primary source of error today is in the samples. Accelerator mass spectrometry (AMS) is the method of choice for ^{14}C dating and it can test very small samples such as a single grain of wheat.

Anomalous dates are not unusual. Samples can become contaminated by carbonates from ground water, bitumen, or some other exogenous source. They may be intrusive and not really belong to the levels where found in archaeological sites. They may be misidentified. The Lewisville site in Texas, for example, was given an early date and thought to be one example of pre-Clovis occupation by Native Americans. On re-examination, it turned out the natives had been burning lignite instead of wood (Dincauze, 1985). In most cases, anomalous dates can be detected by other means; they will be out of context with other evidence. One date for a site is never enough, but if a series of dates is obtained that group well by level and make sense in the context of the site as a whole, the archaeologist can have a good deal of confidence that the excavation has gone well. If the dates range widely and do not make much sense, something is wrong, and the ^{14}C method is probably not to blame. It is a good method of dating, but the samples submitted can be very misleading.

There are other methods of dating, some of which are more accurate than ^{14}C. Tree ring dating can be extremely accurate where a full sequence of patterns is fully worked out. Pot shards can be more accurate than ^{14}C in areas where pottery sequences have been fully developed. Pottery styles, decorations, mode of manufacture, and so on, are remarkably time dependent. Thermoluminescence has been used where other methods are not available. This is based on the theory that heat—such as firing a pot—drives electrons from unstable sites in the clay mineral matrix. Electrons then return to such sites over time. The method is generally less accurate than ^{14}C, but can help where radiocarbon cannot be used. Hydration of obsidian can also give a rough estimate of time. When an obsidian blade is struck from a core, the edges are exposed for the first time since the volcanic glass hardened. The silica hydrates at a fairly constant rate, so that an approximate time since the blade was struck can be estimated. Fission track, amino acid racemization, fluorine diffusion, uranium series, potassium or argon analysis, and other methods are also useful techniques especially for dates before 50,000 years ago. In the far north, lichen growth on rocks has been used with some success for certain time ranges, and where the yearly snow melt produces deposition in thin layers of sediment that can be counted like tree rings.

Linguists have developed a system of dating called glottochronology. It involves a taxonomy of languages. Words and syntax change over time in a manner similar to organic species. A change in the sound or use of a word is like a mutation, and a succession of such mutations can often be traced over time. One would suppose that different language groups would change at different rates; linguists have taken this into account. The system has its faults but can contribute useful chronological information.

Dates have been presented with various notations. Following a general convention, I will report dates based on (a) uncorrected ^{14}C estimates as bp/ ad, and (b) either corrected dates or those based on general information as BP/AD. To avoid confusion with bp/BP (before present), I will use the term "years ago." The exception is when referring to millennium BC (before Christ) in this book. BC and BP have sometimes been confounded and one must read the literature carefully.

Archaeological Sequence of Village Sites

Two very important sites, Tell Abu Hureyra and Mureybit, have been excavated (see Figure 8.4). Located near the Euphrates in Syria, they are now submerged under an artificial lake. The lower levels are classified as Epipaleolithic and higher levels are Neolithic so the sites span the time range of interest. Both have yielded abundant carbonized plant remains, sickle blades with sheen, and mortars and grinding stones for processing hard seeds and grains. Among the plant materials were large samples of wild einkorn and wild rye from the lower levels. Abu Hureyra is somewhat the older of the two and ^{14}C dates made on the charred grains themselves by accelerator equipment (AMS) ranged from 11,100 to 9250 BP (Hillman et al., 1989). The cereals are considered to have been harvested from the wild and not cultivated, although they do not occur in the region today.

Further west, along the Jordan rift and near the Mediterranean, the Natufian culture flourished throughout the ninth millennium BC. The lithic industry was first discovered in 1928 by Dorothy A. E. Garrod in Shouqbah Cave and is now known from many sites from Beidha in the southern Jordan highlands near Petra, to basal Jericho, to Mallaha near Lake Houleh, and westward to the coast. Of particular interest is the presence of sickle blades, sickle handles, and even some intact sickles. The blades often have a sheen or gloss, which is taken to indicate they had been used to harvest cereals, either wild or tame. Grinding and pounding equipment, both stationary and movable, was also abundant. At Mt. Carmel, mortars were ground into solid rock. At Mallaha, well-made decorated boulder mortars were found together with plastered storage pits (Perrot, 1966).

Unger-Hamilton (1989) experimentally made a careful comparison of used blades and Natufian blades from several sites. Natufian blades with sheen

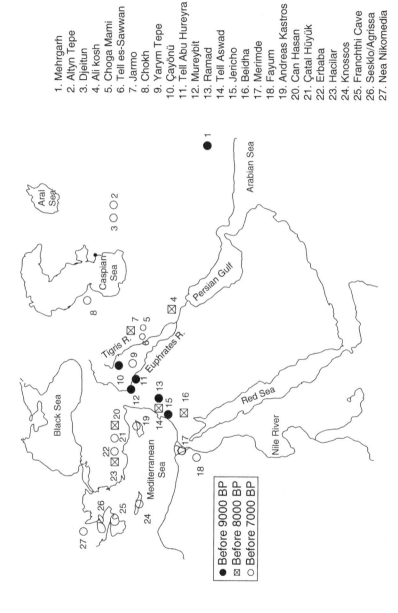

1. Mehrgarh
2. Altyn Tepe
3. Djeitun
4. Ali kosh
5. Choga Mami
6. Tell es-Sawwan
7. Jarmo
8. Chokh
9. Yarym Tepe
10. Çayönü
11. Tell Abu Hureyra
12. Mureybit
13. Ramad
14. Tell Aswad
15. Jericho
16. Beidha
17. Merimde
18. Fayum
19. Andreas Kastros
20. Can Hasan
21. Çatal Hüyük
22. Erbaba
23. Hacilar
24. Knossos
25. Franchthi Cave
26. Sesklo/Agrissa
27. Nea Nikomedia

● Before 9000 BP
⊠ Before 8000 BP
○ Before 7000 BP

Figure 8.4 Approximate dates of some Neolithic farming settlements.

date as early as the ninth millennium BC and the wear patterns show striations similar to those produced experimentally when the plants were growing on loose soil as in a cultivated field. The suggestion was made that the Natufians might have been cultivating cereals by that date. It was also observed, however, that the sheen on the earlier blades matched that from experimental blades used to cut green stem growth. When harvesting wild cereals, it is necessary to cut the stalks before the grain is fully ripe, otherwise the seed is lost to shattering. At this stage, the stalks, especially of wild einkorn are green and juicy at the base. It was not until late Pre-pottery Neolithic B (PPNB) or about 7000 BP that blades consistently had sheen that matched experimental blades used to cut dry straw. The evidence is admittedly tenuous but highly suggestive that the cereals harvested by the Natufians were wild-type in character, whether cultivated or not. Unfortunately, Natufian sites have not yet yielded cereal remains.

All the equipment for cultivating cereal grains is present in the Natufian industries, but there is no indication that either plants or animals were domesticated. The Natufian people lived in an area in which wild wheat and barley are abundant today and presumably were abundant at the time they occupied the area. It may be that natural stands were adequate to supply their needs and cultivation was unnecessary. Harlan and Zohary (1966) raised the question:

> Why should anyone cultivate a cereal where natural stands are as dense as a cultivated field? If wild cereal grasses can be harvested in unlimited quantities, why should anyone bother to till the soil and plant the seed? We suspect that we shall find, when the full story is unfolded, that here and there harvesting of wild cereals lingered on long after some people had learned to farm, and that farming itself may have originated in areas adjacent to, rather than in, the regions of greatest abundance of wild cereals.

There is little point in attempting to pinpoint the very earliest clear evidence of plant or animal domestication. We shall never uncover the real beginnings, and they surely did not evolve in any one place. New excavations will soon make current information obsolete in any case. As of now, we can say that the first traces of cereal cultivation are found in PPNA and date to about 10,000 BP. Barley and emmer appear to be the first crops. PPNA is not abundant, and until recently is known only from the sites of Gesher, Netiv Hagdud, Gilgal, and Jericho all within a radius of 15 km in the Jordan Rift Valley and from Tell Aswad in the Damascus Basin

(Bar-Yosef & Kislev, 1989). PPNB, beginning about a half millennium later is much more abundant. All have emmer, and all but the Çayönü site have barley. The pattern at Abu Hureyra seems to be typical, that is, the epipaleolithic culture terminated about 10,500 BP and was followed about 9500 BP by a fully agricultural PPNB. There was no transition. Once the combination of emmer, barley, sheep, and goats was established, the system spread rapidly throughout the whole Near East. By 9000 BP, an effective system had evolved.

The appearance and spread of crops domesticated in the region can be traced by a study of Table 8.1. For more details, the reader is referred to Zohary and Hopf (1988, p. 249).

In the Near East, animal domestication was closely coordinated with plant domestication. Archaeological sites yield bones in much greater abundance than plant materials, but they also pose problems of interpretation. The earliest domesticates must have been morphologically identical to wild-types, but differences did appear in due time. Goat horns became twisted, and this can be detected in horn cores often found in sites. Cattle, pigs, and sheep became smaller than wild-types. Often there is a shift in the kill to a higher percentage of juveniles and toward more males than females. At Mehrgarh, two graves were found that included an adult human and five kids each. The goats were less than 3 months old. In several sites, a striking shift from gazelle and deer to sheep and goats has been detected. A few sites in or near the shaded area in Figure 8.1 have yielded evidence of domesticated meat animals dated earlier than 9000 BP. At Mehrgarh, goats were domesticated when the village was founded, but sheep and cattle may have been domesticated on the spot independently of activities farther west (Meadow et al., 1984).

The spread of animal husbandry generally coincided with the spread of Neolithic farming, but there are some intriguing features. Sheep appear suddenly in southern and southeastern France in late Mesolithic (in this case sixth millennium BC) well before the arrival of other domesticated animals, cultivated plants, or pottery (Davis, 1987).

The incorporation of animals into agricultural rituals and sacrifice is more than hinted at in some of the sites excavated. Çayönü, excavated for almost three decades (1964–1991) by the Turkish-American team of Robert J. Braidwood and Halet Cambel, revealed some strange and intriguing features. For one thing, the people had been working to some extent with native copper before making pottery. For an early site (ca. 9200 BP) it has remarkably advanced architecture. One building contained a large polished stone slab approximately 2 m × 2 m surrounded by a plastered floor and

Table 8.1 Plant remains from selected early villages. Sources from site reports and Zohary and Hopf (1988). See also Figure 8.1.

Sites[†]	Approx. dates (bp) [‡]	Plants[§]
Tell Abu Hureyra	11,200-10,500	ek, b, wild rye, l
PPNB	9600-8500	ek, EK, em, EM*, b, BN, o, L
PPNA sites (see text)	10,000	EM, B2
Jericho PPNA	10,000-9300	EM, B2*
PPNB	9300-8500	EK, EM*, B2, P, L, F
Tell Aswad	9800-9600	EM*, b, P, L
	9600-8600	EK, EM*, W, b, B2, BN, P, L
Çayönü	9200-8500	ek, EK, em, EM*, b, P, L, V*, F
Ali Kosh	9500?-8750	ek, EK, EM*, b, BN
	8750-8000	ek, EK, EM*, B2, BN, B6, o, L
Beidha	9000	EM*, b, o
Hacilar	8750	ek, EM*, BN, L
Jarmo	8750	em, EM, B2, P, L
Can Hasan	8500	ek, EK, EM, W*, B2, L, V
Tell Ramad	8250-7950	EK, EM*, W, B2, L, F
Tepe Sabz	8000-7000	ek, EK, EM, W, B2*, BN, B6, F
Tell es-Sawwan	8000-7000	EK, EM, W, B2*, BN, B6, F
Choga Mami	8000-7000	ek, EK, EM, W, b, B2, BN, B6, P, L, F
Yarym Tepe	8000-7000	EM, W, B2, BN, B6*, P
Çatal Hüyük	8000-7000	EK, EM*, W, BN, P, V
Erbaba	8000-7000	EK, EM*, W*, B2, BN, P*, L, V
Andreas-Kastros	8000-7000	EK, EM*, B6, L, F
Chokh	8000-7000	EK, EM, W, B2, BN
Mehrgarh	8000-7000	EK, EM, W, B2, BN, B6
Franchthi Cave	8000-7000	EM, B2
Sesklo	8000-7000	EK, EM*, B6, L, V
Nea Nikomedia	8000-7000	EK, EM*, BN, P, L, V
Agrissa	8000-7000	EK, EM, B2, B6, L, millet
Knossos	8000-7000	EM, W*, B2

[†] PPNA and PPNB = Pre-pottery Neolithic A and B.
[‡] Lower case bp are uncorrected ^{14}C date estimates of BP.
[§] lower case = wild; upper case = domesticated; ek, EK = einkorn; em, EM = emmer; W = free threshing wheat; b, B = barley; B2 = 2-rowed; BN = naked; B6 = 6-rowed; o = oats; P = pea; l, L = lentil; V = vetch; F = flax; * = most abundant.

about 100 human skulls, together with some complete and partial human skeletons. Blood residues from the slab were analyzed by Loy and Wood (1989). The traces proved to be blood of human, sheep, and *Bos primigenius* Bojanus, an extinct species of wild cattle. Traces of both wild cattle and human blood were also found on a large black flint knife. Skulls of *B. primigenius* with horns were also found in the building. The authors did not wish to speculate in print on the significance of these findings other than to state the obvious fact that the "skull building" was a site of some kind of mortuary or sacrificial ritual. The size and complexity of this site so early in time may hint at still earlier discoveries to be made with respect to agricultural origins.

There are many details that remain to be worked out, but the general pattern that is coming into focus implies that the nuclear area is rather large and must be taken as a whole unit. Our earliest evidence for both plant and animal domestication falls within the area where wild barley is abundant today and where it may be found in rather primary habitats. The shaded area of Figure 8.1 may serve to outline the nuclear area of early plant and animal domestication in the Near East. It is true that the earliest evidence of domestic cattle falls slightly outside of the region indicated, but the nature of the site shows clearly that agriculture had been well established before the founding of Çatal Hüyük.

The nuclear area must be understood as a unit because events in one part of the area were not without influence on other parts. Agriculture did not arise full-blown in one spot but as a synthesis of practices and techniques that had different origins. Some people concentrated on sheep and goats, others on pigs or cattle. Emmer, barley, and einkorn were probably introduced into cultivation by different people in different subregions of the area and at different times. There were probably multiple domestications of both plants and animals. It was the integration of several practices and techniques that established food production as an ecological system.

One concrete evidence of contact across the nuclear area is the distribution of obsidian tools, mostly blades. Within this region, obsidian occurs only near Lake Van and central Anatolia in Turkey. By 9500 BP there is evidence for an obsidian trade extending throughout the nuclear area (Wright & Gordus, 1969). People of Beidha or Ali Kosh had some sort of contact with the Lake Van area to obtain the obsidian found in these sites. The trade reached preagricultural sites as well. By the early seventh millennium BC there might even have been some trade in native copper.

Spread of Agriculture Out of the Nuclear Area

The spread of agriculture out of the nuclear zone has been charted in detail by Zohary and Hopf (1988, p. 249). They plotted a select sample of 100 sites with approximate dates and plant remains reported for each. Anyone interested in the diffusion of individual crops from southwest Asia across Europe and the Mediterranean region should consult this excellent review. Figure 8.4 represents a simplified version showing approximate dates of some early farming settlements. A temporal sequence is clearly indicated. The nature of diffusion is under debate, however. Ammerman and Cavalli-Sforza (1984, p. 175) argue for a wave front of farmers advancing across Europe at approximately 1 km per year. Barker (1985), Dennell (1983), Gregg (1988), and others disagree, insisting that the temporal sequence was due to progression of a mosaic or patchwork of interactions between Neolithic and Mesolithic cultures.

The wave of migration theory gets support from plotting dated sites of both Neolithic farming settlements and terminal Mesolithic occupations. The distribution of human gene alleles is also suggestive (Ammerman & Cavalli-Sforza, 1984, p. 175). The opposing view points to the fact that the early Neolithic settlers in each region occupied small scattered sites on loess soils, avoiding the clayey soils. Mixed assemblages indicate a considerable period of trade and interaction between the two cultures. The diffusion was also erratic and jumpy, and the jumps seem to correlate with changes in climate. The pattern does not fit that of a tidal wave of farmers. Barker (1985, p. 255) sums up the mosaic development views this way: "For virtually all of Europe, therefore it seems to me most likely that the various systems of initial farming which we can discern were developed by the indigenous populations rather than by newcomers."

Both sides conceded that the spread of agriculture across Europe could have been due to a mixture of developments. Some farmers probably did migrate, for example, the Bandkeramik people. Angel (1984, p. 53), based on skeletal evidence, states: "Early farming populations of the seventh and sixth millennium BC at Çatal Hüyük, Nea Nikomedia, Franchthi Cave, and Lerna, for example, include descendants of Semitic-speaking Africans and of Balkan riverine populations." No doubt some Mesolithic people took up agriculture and others opted to come to terms with their farming neighbors. Farmers, no doubt, increased their own hunting–gathering activities when crops were bad; poor yields were probably common. None of the suite of crops from the winter rainfall–summer drought zone could have been initially well adapted to central and northern Europe. The different food procurement systems are not mutually exclusive, nor does the development of European agriculture depend on one or the other mode of diffusion. The cereals also moved eastward. Djeitun and Altyn Tepe on the Iran border date to sixth millennium BC and had emmer and barley. Mehrgarh, in Baluchistan, is probably a little older and reports einkorn, emmer, free-threshing wheat, and barley. We do not know when agriculture reached Pakistan and India, but the Mohenjo-daro and Harappa civilizations seem to have been based on a wheat and barley culture. They arose in the early and middle third millennium BC. Wheat reached China by the late 5000–6000 years BP (Betts et al., 2014), but barley arrived sometime later by about 4000 years BP (Ho, 1969).

The Near Eastern crop complex also reached the Ethiopian highlands at a fairly early time, but there is as yet no primary evidence to estimate the date. The Near Eastern crops have been in Ethiopia long enough to develop centers of diversity (Harlan, 1969) and to have evolved unique varieties that are not found elsewhere. The emmers of Ethiopia, however, are more closely related to those of southern India than to those of Europe.

Recorded History

The historic time range is too late to tell us much about the origin of the primary crops and agricultural development, but it is not without interest. In the Mesopotamian Valley, where writing appears to have originated, barley was the primary crop. In the south especially, barley almost completely displaced wheat as the cereal crop by 4300 BP. This is taken to be not so much from preference as from necessity since there is independent evidence that the irrigated lands were salting up. Barley is much more salt-tolerant than wheat. Even before this shift to a near monoculture, however, barley was apparently the more important of the two (Adams, 1965).

Barley also held a dominant position even into classical Greek times. It was the food of the poor and the ration of the soldier. It is not an attractive cereal from a dietary and culinary point of view. The covered sorts, especially, are very high in fiber and difficult to digest. The culture of naked varieties improves the diet considerably, but even naked barley is less desirable than wheat. From Table 8.1, we can see that naked barley appeared rather early and spread quickly. Today it is found only where barley is a significant part of the human diet.

In the cuneiform literature of Mesopotamia, barley is generally mentioned much more often than wheat. There is a myth concerning the divine origin of barley but not a corresponding one for wheat. The relative value as indicated by price, taxes, or rations shows barley and emmer (probably in the glume) about equal and about half the value of naked wheat. Yields at about 4400 BP were calculated by Adams (1965) from a number of records and show: barley, 2537 L/ha; emmer, 3672 L/ha; and wheat, 1900 L/ha. The emmer was presumably in the glume and, therefore about 75% as heavy as wheat. The yields are quite respectable but similar computations only a few centuries later indicate a sharp decline that, again, may be attributed to salinization. By 4100 BP the yield of barley was only 1460 L/ha and wheat had virtually disappeared as a crop in the southern region (Jacobsen & Adams, 1958).

In Egypt, where salting was less of a problem, emmer was the preferred cereal for bread, and naked free-threshing types were not grown until the Greek occupation after Alexander the Great. When Herodotus visited Egypt in the fifth century BC, he wrote that the Egyptians ate emmer and considered it a disgrace to eat (naked) wheat. Under Greek influence, however, they did change to bread wheat after a big export market opened up in Rome.

Einkorn apparently never reached Egypt, Ethiopia, or India. It moved with the early agriculturalists up the Danube and into Western Europe, but seems to have been a relatively minor crop almost everywhere it was introduced. Records indicate it was important to Schwabia (region in SW Germany) and is still grown in mountainous regions of Switzerland, Italy, and Germany. In recent years, it has been reportedly grown on a small scale in France, Spain, Morocco, and the Balkans. The only place that it is grown on a large scale today is in Turkish Thrace where it is used as livestock feed.

Conclusions

Long before recorded history, the people of the Near East had become completely and utterly dependent on agriculture for their food. There was no possible way to return to gathering economies. The threat of famine was ever at hand. If the rains failed, starvation stalked the land. Biblical literature makes repeated reference to years when the heavens were shut up and the rains did not come. While the Israelites were still pastoralists, they were forced by drought to move to Egypt, but even Egypt could suffer hard times. There is evidence that the Old Kingdom went into eclipse because of a series of years with low floods. The inscription on the tomb of Ankhtifi was written during this first "dark age" of Egypt when the Old Kingdom came to a close and before the rise of the Middle Kingdom. The full text is starkly eloquent.

Another commentary of the time is the lament of the Egyptian prophet Ipuwer, which reads, in part (Erman, 1927):

Plague stalketh through the land and blood is everywhere. . . . Many men are buried in the river . . . the towns are destroyed and Upper Egypt is become an empty waste . . . the crocodiles are glutted with what they have carried off. Men go to them of their own accord. Men are few. He that layeth his brother in the ground is everywhere to be seen . . . grain hath perished everywhere . . . the storehouse is bare, and he that hath kept it lieth stretched out on the ground.

What profited man to domesticate barley, wheat, sheep, goats? What was gained by the development of effective food-producing systems? Obviously, it was not an assured or stable food supply. However, when the system works well, large numbers of people can be supported and civilizations can emerge from an agricultural base. The pyramids were all built before Egypt's first Dark Age, and the splendor, wealth, and power of the Old Kingdom were unmatched anywhere in the world in early third millennium BC. The ancient civilizations of the world are the visible fruits of the evolution of agricultural economies.

References

Adams, R. M. (1965). *The land behind Baghdad: A history of settlement on the Diyala plain.* Chicago, IL: University of Chicago Press.

Ammerman, A. J., & Cavalli-Sforza, L. L. (1984). *The neolithic transition and the genetics of populations in Europe.* Princeton, NJ: Princeton University Press. https://doi.org/10.1515/9781400853113

Angel, J. L. (1984). Health as a crucial factor in the changes from hunting to developed farming in the eastern Mediterranean. In M. N. Cohen & G. J. Armelagos (Eds.), *Paleopathology at the origins of agriculture* (pp. 51–73). New York, NY: Academic Press.

Barker, G. (1985). *Prehistoric farming in Europe.* Cambridge, UK: Cambridge University Press.

Bar-Yosef, O., & Kislev, M. E. (1989). Early farming communities in the Jordan valley. In D. R. Harris & G. C. Hillman (Eds.), *Foraging and farming: The evolution of plant exploitation* (pp. 632–642). London: Unwin Hyman.

Bell, B. (1971). The dark ages of ancient history. I. The first dark age of Egypt. *American Journal of Archaeology, 75,* 1–26. https://doi.org/10.2307/503678

Betts, A., Jia, P. W., & Dodson, J. (2014). The origins of wheat in China and potential pathways for its introduction: A review. *Quaternary International, 348,* 158–168. https://doi.org/10.1016/j.quaint.2013.07.044

Davis, J. M. (1987). *The archaeology of animals*. New Haven, CT: Yale University Press.

Dennell, R. (1983). *European economic prehistory: A new approach*. New York, NY: Academic Press.

Dincauze, D. F. (1985). An archaeological evaluation of the case for pre-Clovis occupations. *Advances in World Archaeology, 3*, 275–323.

Erman, A. (1927). *The literature of the ancient Egyptians*. New York, NY: E.P. Dutton and Company.

Gregg, S. A. (1988). *Foragers and farmers: Population interaction and agricultural expansion in prehistoric Europe*. Chicago, IL: University of Chicago Press.

Harlan, J. R. (1967). A wild wheat harvest in Turkey. *Archaeology, 20*, 197–201.

Harlan, J. R. (1969). Ethiopia: A center of diversity. *Economic Botany, 23*, 309–314. https://doi.org/10.1007/BF02860676

Harlan, J. R., & Zohary, D. (1966). Distribution of wild wheats and barley. *Science, 153*, 1074–1080. https://doi.org/10.1126/science.153.3740.1074

Hillman, G., Colledge, S. M., & Harris, D. R. (1989). Plant-food economy during the epipalaeolithic period at Tell Abu Hureyra, Syria: Dietary diversity, seasonality and modes of exploitation. In D. R. Harris & G. C. Hillman (Eds.), *Foraging and farming. The evolution of plant exploitation* (pp. 240–268). London: Unwin Hyman.

Ho, P.-T. (1969). The loess and the origin of Chinese agriculture. *The American Historical Review, 75*, 1–36. https://doi.org/10.2307/1841914

Jacobsen, T., & Adams, R. M. (1958). Salt and silt in ancient Mesopotamian agriculture. *Science, 128*, 1251–1258. https://doi.org/10.1126/science.128.3334.1251

Langdon, S. H., & MacCulloch, J. A. (1931). *Semitic. The mythology of all races*. (*Vol. 5*). Boston, MA: Marshall Jones Co.

Loy, T. H., & Wood, A. R. (1989). Blood residue analysis at Çayonü Tepesi. *Journal of Field Archaeology, 16*, 451–460.

Meadow, R. H., Clutton-Brock, J., & Grigson, C. (1984). Animal domestication in the Middle East: A view from the eastern margin. *Animals and Archaeology, 3*, 309–337.

Perrot, J. (1966). Le gisement Natoufien de Mallaha (Eynan), Israel. *L' Anthropologie, 70*, 437–484.

Unger-Hamilton, R. (1989). The epi-palaeolithk southern Levant and the origins of cultivation. *Current Anthropology, 30*, 88–103. https://doi.org/10.1086/203718

Van Zeist, W., & Dottema, S. (1982). Vegetational history of the eastern Mediterranean and the Near East during the last 20,000 years. In J. L. Bintiff & W. Van Zeist (Eds.), *Palaeoclimate, palaeoenvironments and human communities in the eastern Mediterranean region in later prehistory, British Archeological Reports International Series no. 133 (i and ii)* (pp. 277–302). Oxford, UK: British Archaeological Reports.

Wright, G. A., & Gordus, A. A. (1969). Distribution and utilization of obsidian from Lake Van sources between 7500 and 3500 B.C. *American Journal of Archaeology, 73,* 75–77. https://doi.org/10.2307/503380

Zohary, D., & Hopf, M. (1988). *Domestication of plants in the Old World.* Oxford, UK: Clarendon Press.

9

Indigenous African Agriculture

I have given thee Punt. No one knows the way to the Land of the Gods anymore; no one has gone up to the terraces of incense, none among the Egyptians. They have only heard tales of olden times repeated by word of mouth.

Oracle of Amon, approximately 3500 BP
(Doresse, 1957; J. R. Harlan translation)

Where the south declines toward the setting sun lies the country called Ethiopia, the last inhabited land in that direction. There gold is obtained in great plenty, huge elephants abound, with wild trees of all sorts, and ebony; and the men are taller, handsomer, and longer lived than anywhere else. Now these are the furthest regions of the world.

Herodotus, 2447 BP (as recorded in Komroff, 1928)

Source: Patricia J. Scullion

Harlan's Crops and Man: People, Plants and Their Domestication, Third Edition.
H. Thomas Stalker, Marilyn L. Warburton, and Jack R. Harlan.
© 2021 American Society of Agronomy, Inc. and Crop Science Society of America, Inc.
Published 2021 by John Wiley & Sons, Inc.
doi:10.2135/harlancrops

Introduction

While the developments we have recorded were taking place in the Near East, something was going on in Africa south of the Sahara, but we know little about it. Contact between the Mediterranean world and sub-Saharan Africa was extraordinarily tenuous from the beginnings of recorded history until the rise of Islam, and the Western World did not learn much of Africa until Portuguese explorations in the 15th century. Among the masses of beautifully preserved plant materials found in ancient Egyptian tombs there is not a single indigenous African crop, although some of the wild native plants were collected. Egypt and the whole of North Africa looked northward; their culture and their agriculture belonged to the Mediterranean world.

An indigenous agriculture was developed in Africa, by Africans domesticating African plants. Domestication in Africa started with animals, and moved much later to crops, at about 5000 BP in the Sahara and Sahel. These areas were wetter at the time, and this allowed the domestication of sorghum (Winchell et al., 2017) and pearl millet (Manning et al., 2011). After the Sahara and Sahel dried in 4000 BP, agriculture shifted to wetter climates further south (Ehret, 1998). An agricultural system evolved with a farming village pattern and spread over much of the continent. It was adequate to support the high cultures of Mali, Ghana, Nok, Ife, and Benin. The system was complete with cereals, pulses, root and tuber crops, oil crops, vegetables, stimulants, medicinal crops, and magic and ritual plants.

The number of plants domesticated is impressive (Table 3.1). The most important of these from a worldview are coffee, sorghum, pearl millet, oil palm, watermelon, cowpea, and finger millet. The most important to the Africans as food are sorghum, pearl millet, African rice, yams, oil palm, karité, cowpea, bottle gourd, finger millet, tef, enset, and noog. Other crops of considerable importance to the Africans are fonio, cola, chat, okra, roselle, *Vigna*, and *Corchorus olitorius* Many indigenous African crops have been replaced by higher yielding and more easily cultivated crops from other regions, especially maize, cassava, and banana (M'Mbogori, 2017).

Archaeological Prelude

The prehistory of man extends further back in time in Africa than anywhere else on earth. The evidence, as we now understand it, indicates that the genus *Homo* originated in Africa well over two million years ago, and most of human evolution has taken place in this arena. In this sense we are all Africans, and Africa is the home of the human race, but studies of our own origins are concerned with a time range far too early to be of help in understanding the origins of African agriculture.

The time range of concern to us begins with the terminal phase of the glaciation in Europe, about 11,000–12,000 years ago. Evidence for changes in climate in Africa is not as subtle as those in the Near East. Any observant amateur can recognize stabilized dune formations in the broad-leaved savanna regions both north and south of the forest zones. Fossil lake shorelines and fossil streams are conspicuous. The levels of lakes Chad, Rudolph (Turkana), Afrera, Nakuru, Naivasha, Magadi, Victoria, and Katwa changed rather radically within this time span. Lake Chad was once 10 times its present size, and Lake Rudolph was once so deep that it overflowed into the Nile watershed (Leblanc et al., 2006; Morrissey & Scholz, 2014). A series of terraces along the Nile River indicate a succession of rather spectacular rises and declines of the river level.

Several surveys that have been made involving geology, hydrology, and palynology all agree that the changes in climatic patterns have been complex and difficult to interpret. Rises and declines in the lake levels have not been synchronous; long-term trends were interrupted by short-term countertrends. The details are too complex to deal with here, but the most general shifts in climate were sketched by Clark (1967) and Muzzolini et al. (1989).

Africa, at the end of the Pleistocene, say 14,000 BP, was in a hyperarid mode. Forests had retreated to rather small refuges along the Atlantic coasts and parts of the eastern highlands. The Sahara was virtually uninhabited. Along the Nile, some people developed a "Nilotic adaptation." They were big game hunters with wild cattle one of their favorite foods. They also exploited aquatic resources and caught a lot of fish. This was probably assisted by "wild" flooding of the Nile during an episode 13,000–12,000 BP. The waters spilled over the alluvial plain and left fish stranded in shallow pools when they receded. Resources were sufficiently abundant so that people were more or less sedentary and occupied the same sites for long periods of time. The stone tool kit was Epipaleolithic, featuring microliths and small blades, but at several sites there were also grinding stones, including heavy mortars and blades with sickle sheen. The equipment suggests heavy use of wild grass seed harvests.

The Nilotic adaptation suggested to C. O. Sauer that sedentary fishermen would be the most likely people to start plant domestication. Nevertheless, these cultures began to fade as increasing rainfall brought improved conditions to the Sahara. A pluvial that set in about 12,000 BP peaked from 9000–8000 BP and the number of sites in the Nile Valley declined; those in the Sahara increased. This wet phase was abruptly terminated by a short arid phase peaking out about 7000 BP which, in turn, was relieved by a "Neolithic pluvial" some 6500–5000 BP when there was more rainfall than in modern times. Desiccation set in again, reaching more or less the present rainfall levels by about 4500 BP (Butzer et al., 1989).

The critical period of interest, then, is that between the hunter-gatherers of the Nilotic adaptation and the fully developed farmers. By 7000 BP, farmers in the Nile Valley were irrigating their crops (Conniff et al., 2012) and fully developed agriculture was practiced. However, it does seem certain that the first African agriculture developed somewhere other than the Nile Valley.

Farther west in North Africa, we find traces of hunter-gatherers who hunted large animals and sometimes occupied sites off and on for millennia, even if not fully sedentary. Some groups specialized in hunting certain species; some specialized in wild sheep (in this case, *Ammotragus* rather than *Ovis*), others in aurochs (*Bos primigenius*), others various antelope, and so on. They had a few grindstones. The best known and widespread of the North African cultures is the Capsian. It developed in the early Holocene and lingered on here and there to about 6000 BP. The stone tool assemblage was Epipaleolithic throughout, but demonstrated an evolution toward smaller and more finely made microliths and blades. There is no convincing evidence of either plant or animal domestication. Capsians were noted for their fondness for snails and large middens of shells are found at archaeological sites. This was probably not a major part of the diet, however, because they were highly seasonal in availability.

Some investigators have advanced arguments for herding of sheep, goat, cattle, and even antelope based on one or more of the following lines of evidence: (a) a high proportion of bones from young animals; (b) reduced size, especially of cattle; and (c) the argument that the progenitors of sheep and goat were not present in Africa at that time and, therefore, caprine bones must represent domestic races. Each of the arguments has flaws. A high percentage of young animals could indicate a technique of selective culling without actual herding. The reduced size of aurochs could have been due to genetic response to desiccation, and we are not altogether sure of the distribution of wild sheep and bezoar goat at that time range. Davis (1987), following Isaac (1970), shows a distribution for them in the Near East and not in Africa. Muzzolini et al. (1989) was not so sure. Past distributions might be very different from present ones. Furthermore, it is difficult to tell sheep from goat unless diagnostic bones or horn cores are available and distinguishing between *Ovis* and *Ammotragus* is not much easier.

By whatever means or by whatever route, nomadic pastoral economies were established across the Sahara during the Neolithic pluvial of 6500–5000 BP. The archaeological evidence seems to suggest local developments rather than invasions from outside. Livestock tending and ceramic preparation were developed in the Sahara by people with an Epipaleolithic tool inventory well before ceramics and domestic animals were known in the Nile Valley.

Indeed, the Sahara began to be reoccupied around 10,000 BP, with people living in the western desert of Egypt at sites like Abu Ballas, El Adam, and Nabta Playa (Figure 9.1). People reoccupied the Air by 9400 BP (Tagalagal, Ardar Bous) and the Hoggar by 9000 BP. These groups also had Epipaleolithic tool assemblages, plus abundant grinding equipment, blades with gloss, and pottery. The pottery was fragile and rather poorly made, but occurred much earlier than any in the Nile Valley or Mediterranean North Africa. These people often camped on the shores of playa lakes that expanded in the rainy seasons and contracted in the dry seasons. They exploited aquatic resources, hunted upland game, and probably harvested wild grass seeds. The lakes and streams present in those days have now dried up.

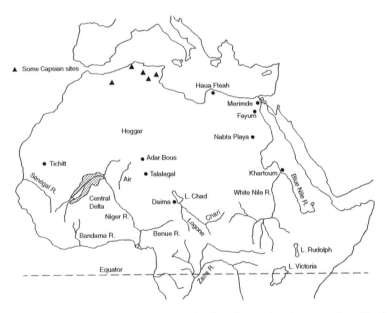

Figure 9.1 Map of Northern African showing sites and regions mentioned in the text. The sites of Abu Balles and El Adam are too close to Nabta Playa to fit on the map.

Trends toward sedentary living continued, and at Nabta Playa some 14 circular houses were found dating to about 8000 BP and arranged in two rows as if along a street. Toward the end of the main pluvial, approximately 7500 BP, a widespread assemblage of cultures had become established ranging from Mali in West Africa to the Ethiopian plateau. The cultures were not necessarily related genetically but had in common adaptation for exploiting water resources (fish, turtle, shellfish, hippopotamus, etc.). Pottery of "Early Khartoum" or "Khartoum Neolithic" types with "wavy line" or comb-impressed decoration was widespread throughout the region. Harpoons and grinding stones were conspicuous in the tool assemblages. Here and there true villages appeared with houses of mud wall construction. A sedentary way of life had evolved, but with no evidence of plant domestication.

About 6000 BP, fully developed farming systems began to appear in the Nile Valley. The earliest so far excavated are Merimde on a branch of the Nile in the delta and Fayum on the shores of a lake fed by the river some distance upstream from the delta. Merimde covers some 18 ha, has sun-dried mud brick houses arranged in streets, plastered pits, and large ceramic jars for grain storage. Bones of domestic cattle, pigs, dogs, sheep, goat, and donkey were found. The sheep were larger than those of the Old Kingdom to come later. Fully domesticated crops included emmer, covered six-rowed barley, lentil, peas, and flax, but the cereals were by far the most abundant. At Fayum, granaries were found containing emmer, both two- and six-rowed barley, and some flax. Within a few centuries, many sites of a full Neolithic economy had spread up and down the Nile Valley and Egypt was prepared to move into the Old Kingdom and historical times. However, Northern African agriculture presents the curious fact that agriculture arrived along the Nile fully developed by people with an African stone tool assemblage and a suite of crops of Near Eastern origin. Of the animals, cattle, pigs, and donkey could have been domesticated independently in Africa; the sheep and goats may or may not have been depending on mid-Holocene distribution of the progenitors. True indigenous African plant domestication first occurred in the green Sahara, shortly before agriculture moved south during the desiccation after 6500 BP.

Fundamental to Egyptian agriculture is the fact that the Nile floods in late summer. Planting must, therefore, be in the fall, and only cool-season crops can be used. The traditional crops of ancient Egypt were barley, emmer, flax, chickpea, pea, lentil, lettuce, broad bean, onion, leek, and so on, all cool-season species. There were, in addition, perennial crops such as fig, sycamore fig, grape, olive, pomegranate, and others. However, indigenous, warm-season African crops could not be used in the natural flood area. Only when water-lifting devices like the shaduf, water wheel, the screw of Archimedes, and so on, were available could crops be grown in spring and summer. Some indigenous African crops were then introduced, but the shaduf did not arrive until late in the second millennium BC, and the other devices were still later. All required a great deal of labor to irrigate small areas of land, and these tended to be devoted to garden produce rather than field crops.

The linguistic evidence consistently indicates a considerable antiquity for agriculture in sub-Saharan Africa. Tracing the evolution and distribution of words for plow, cow, wheat, barley, and the like, however, Ehret (1984) concluded that agriculture had been established in the Horn of Africa by 7000 BP and enset cultivation was practiced in Ethiopia at roughly the same time (Ehret, 1979). Using words associated with indigenous African agriculture (such as pearl millet, sorghum, flour, porridge, as well as sheep, goat, cow, and the like) he concluded agriculture was practiced by 8000 BP and around Lake Chad by 6000 BP in West Africa (Ehret, 1984).

An early example of plant remains of an African crop were seeds of finger millet found in a rock shelter in Ethiopia (Phillipson, 1977). The dating was not very precise, but they could have been as early as the fourth millennium BC based on the context in which they were found. The identification was made in the University of Illinois Crop Evolution Laboratory and the seeds were clearly of domesticated finger millet, but were not carbonized and looked so fresh that I wondered if some worker on the dig had a hole in his pocket. Phillipson later submitted a sample for AMS analysis to the Oxford Laboratory. They turned out to be a little over 800 years old (Gowlett et al., 1987). It should be emphasized that this was not due to careless excavation. Phillipson had every right to believe the seeds belonged to the contexts in which he found them, but the example does illustrate problems with intrusive material. This is far from the only case; other excavators have had the same problem. Munson et al. (1976) found evidence for domesticated pearl millet replacing wild grass seed harvesting in Mauritania around 3000 BP.

Newer and more comprehensive archaeological studies have found sorghum domestication occurred 5000 years BP in present-day Sudan near the Atbara and Gash rivers (Fuller & Stevens, 2018; Winchell et al., 2017). Sorghum had reached India around 4000 BP, which agrees with the newer studies. Sites in northern Africa have not yielded remains of African crops and sites south of the Sahara appear to be later, for example, sorghum appears to have reached West Africa after 3,000 BP (Clark & Brandt, 1984; Close, 1988; Fuller & Stevens, 2018). Therefore, we must depend more on evidence from the plants and their genes, and this research is now becoming available.

A Savanna Complex

Unlike the Near Eastern agricultural complex, African crops lack cohesion. Many have very limited distributions; some are found only in Ethiopia, some only in limited areas of West Africa. There is no apparent center of plant domestication and activities of domestication seem to have ranged over a vast area from the Atlantic to the Indian Ocean south of the Sahara and north of the equator. African agriculture is, however, basically a savanna agriculture. Sorghum is a savanna crop, not well suited to the high rainfall of the forest zones. Pearl millet is one of the most drought-resistant of all the crops and becomes the dominant one near the fringes of the Sahara (Figure 9.2).

African rice was domesticated from an annual wild rice, *Oryza barthii* A. Chev. (syn.: *Oryza glaberrima* subsp. *barthii*) which is a plant of water holes in the savanna zone (Figure 9.3). In fact, the truly wild forms do not appear in the forest zones or in the derived savanna today, even though a derived weed type does infest rice fields in the high rainfall belt. The *Vigna* or Bambara groundnut is a highly drought-resistant savanna plant. Even the yams, which are the staff of life of the tribes of the forest belt from central Ivory Coast to Cameroon, are basically savanna plants. Their large tubers are adaptations for storage enabling them to survive the dry season and periodic burning.

Karitè, baobab, tamarind, and *Parkia* are all savanna trees. The oil palm was originally a tree of forest margins since it is not tolerant of deep shade. It has spread into the forest and thrives under the disturbance of shifting cultivation, but this is not its natural habitat. Wild cowpeas and hyacinth beans are also forest margin plants. Fonio, roselle, *C. olitorius*, bottle gourd, and watermelon are all savanna plants. Tef, enset, noog, and finger millet are plants of the cool East African highlands. The only true forest plants of the African agricultural complex are cola, coffee, malaguette, and a few other minor plants.

The African savanna complex did spread out of its hearth in a way similar to the Near East agricultural complex. The system spread southward along either side of the East African Rift into southern Africa. Sorghum, pearl millet, finger millet, cowpea, and hyacinth bean went to India and became

Figure 9.2 Known distribution of wild pearl millet (*Pennisetum glaucum* subsp. *violaceum*), solid circles. The northern pearl millet belt is shown by heavy shading. The crop is grown in the cross-hatched zone, but sorghum is more important there. There is a southern millet zone in association with the South African deserts, but it is less clearly defined.

very important there. Guar is wild in Africa but a crop in India. Roselle and *C. olitorius* are vegetables in Africa but fiber plants in India. Noog is grown on a small scale in India but may have been introduced rather recently. The most important of these African plants that went to the Indians are the drought-resistant cereals—sorghum and pearl millet; they are the staff of life for millions in the drier sectors of India. Archaeological finds in India suggest that these crops may have arrived in the second millennium BC (Agrawal, 1982). There are Sanskrit words for pearl millet and finger millet but not for sorghum.

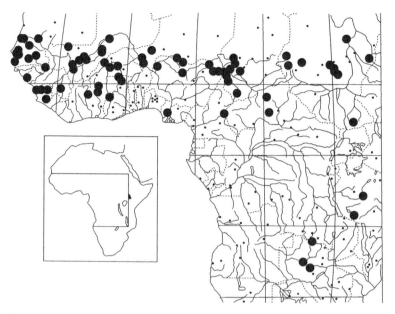

Figure 9.3 Known distribution of wild African rice (*Oryza glaberrima* subsp. *barthii*), solid circles. The grass is native to the savanna, and colonies within the forest zone appear to be recent introductions.

Much of the evidence we have led me to conclude that agriculture in Africa is basically noncentric, that plants were domesticated throughout a broad zone from the Atlantic to the Indian Ocean and primarily within the Sahel and Guinea savanna zones (Figure 9.4). However, new genetic data indicate that for some species, a distinct center can be identified. For example, genetic sequencing of pearl millet indicates that it was domesticated in the Western Sahara 4900 years ago; however, frequent gene flow from an extensive wild species population made it difficult to pinpoint this before extensive sequencing became available (Burgarella et al., 2018). Once established in the savanna, it slowly encroached into the forest through practices that often produced derived savanna. Parts of the complex spread into East Africa and reached India by 4000 years ago. When this all began is still a matter of conjecture. The grinding stones along the Nile and out into the Sahara are a common tool of hunter-gatherers and are still used by Australian Aborigines today. The sickle sheen implies the use of small grains like barley or wild grass seeds rather than sorghum or pearl millet. If a winter rainfall regime prevailed at the time, this is only reasonable. The Qadan culture (15,000–10,000 BP) of the present-day south Egypt region were primarily hunters but cared for and harvested wild grains, including barley; archaeological evidence reports sickles and

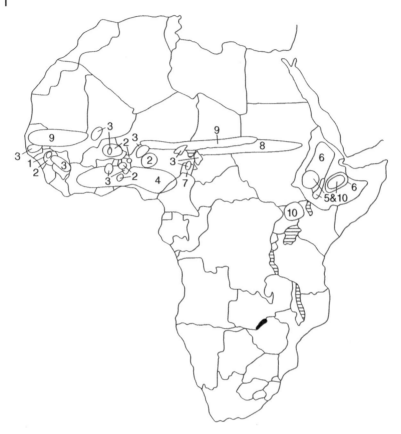

Figure 9.4 Probable areas of domestication of selected African crops:
(1) *Urachloa deflexa*, (2) *Digitaria exilis* and *Digitaria iburua*, (3) *Oryza glaberrima*,
(4) *Dioscorea cayenensis* subsp. *rotundata* (Poir.) J. Miège (syn. *D. rotundata*), (5) *Ensete ventricosum* (Welm.) Cheesman and *Guizotia abyssinica*, (6) *Eragrostis tef*, (7) *Voandzeia* and *Kerstingiella*, (8) *Sorghum bicolor*, (9) *Cenchrus americanus* (L.) Morrone [syn.: *Pennisetum glacum* (L.) R.Br.], and (10) *Eleusine coracana*. *Source:* Reprinted from Harlan (1971). Reprinted with permission from AAAS.

grinding stones (Wendorf et al., 1979). Barley pollen has been tentatively identified in the Sahara phase along the Nile and stands of wild barley occur today in Cyrenaica and Morocco. It appears that the zone of wild barley harvesting may have once been much larger than we had thought, but as of now, we have no evidence of barley domestication in Africa.

When, then, were the African plants domesticated? New sequencing data give us the first concrete evidence. As first suggested, the process went on over a period of several millennia. Some indigenous agriculture was being practiced by 6000 BP or earlier.

Crop Competition and Distribution

The distribution of African crops and cultures may suggest a sequence of events even if the exact dates of the events are not known. The very sharp demarcation between rice-eating tribes and yam-eating tribes has attracted much attention. In the Ivory Coast, people on the right bank of the Bandama River eat rice; people on the left bank eat yams (Figure 7.2). Each crop is very deeply enmeshed in the culture of the people. Rice-eaters do not feel they have eaten a meal unless rice is served. Yams are central not only in the diets of yam-eating tribes, but also in their ceremonies, rituals, myths, and folklore (Coursey, 1972).

The most reasonable explanation for the very sharp separation of cultures appears to involve a series of events. African rice had to be domesticated in the savanna zone, since that is where the progenitor is found (Figure 9.3). Cultivars were developed that could be grown under high rainfall on upland areas in the forest zone. This permitted the rice-eating tribes to expand to the Guinea coast along the Atlantic Ocean. Rice cultures tend to be expansive, but when these people reached the yam belt they found an agricultural system already established. In fact, the social structures of the yam cultures are very strong and more systematically organized than those of other tribes in West Africa. The implication, then, is that yam agriculture and rice agriculture had independent origins and that the yam cultures may have been the older (Coursey, 1972). At least the yam cultures were well entrenched when the rice cultures arrived in the adjacent forest zone.

In Ethiopia, evidence of crop–culture competition takes on other forms. The bulk of Ethiopian agriculture is based on the Near East complex of barley, wheat, grain legumes, flax, safflower, and so on. This imported agriculture apparently met an indigenous system already in place when it arrived. The locally domesticated crops are tef, noog, enset, finger millet, chat, coffee, and so on. The two systems have blended so that now tef, wheat, maize, sorghum, and barley are all grown extensively (Taffesse et al., 2012). Noog is a widely grown oil crop over most of the country, and enset is a staple and the staff of life for a number of tribes in the central and southern highlands. There is a genuine enset culture distinct from those that depend on seed crops.

Finger millet is largely grown in Ethiopia and Uganda, but has spread westward to Lake Chad and a little beyond. There it encountered the West African millet, fonio, and went no further. Other examples could be cited, but it seems evident that crop distributions in Africa depend more on the distribution of tribes and cultures than on the ecological adaptation of the plants.

The introduction of crops from the Americas has had a profound impact on African agriculture. Manioc (cassava) is now more important in Africa than in the New World where it originated. Maize has displaced a great deal of the sorghum acreage in South and East Africa. Capsicum peppers are absolutely basic to much of the African cuisine. Other American crops that have become important are sweet potato, cotton, peanut, tomato, papaya, and tobacco. All have been received well and have competed successfully with indigenous counterparts.

Recorded History

Written history is, of course, too late to tell us anything about agricultural origins, but it is worthwhile to call attention to one event recorded about 3500 BP. During the 18th dynasty of Egypt, Queen Hatshepsut sent an expedition to Punt, thought to be somewhere on the horn of Africa. It was the first government-sponsored plant exploration expedition in recorded history. She had built a temple at Deir El Bahari and wished to establish incense trees on the terraces (Naville, 1898, p. 776). Five ships were dispatched to Punt and were greeted by inhabitants. Scenes on the temple walls show round, thatched houses on stilts with ladders reaching to the doors. The people had cattle and donkeys and were able to supply goods in quantity (Naville, 1898). Incense trees were potted into huge tubs and loaded on board. The cargo from Punt included much incense, gums, resins, ivory, jewels, metal rings, hides, leopard skins, ebony, and other items. Despite serious defacing and vandalism of the temple, a fascinating glimpse is presented of sub-Saharan Africa in the middle of the second millennium BC (Figure 9.5). It shows as well that plant exploration and introduction are venerable activities and can receive governmental support from time to time.

There is a body of literature concerning wild grass seed harvesting that seems to have escaped the notice of anthropologists. Much of it was reviewed by Harlan (1989). Wild grass seeds were a major portion of the diet of people in the Sahara and in some sections of sub-Saharan Africa as late as the 19th century and early 20th century. The scale of the harvests was impressive. Travelers reported warehouses stacked with sacks of it and camel caravans taking loads of wild grass grains from areas of surplus to areas of deficit. One of the desert grasses was noted to sell for one-third the price of barley in the market and not because it was a less desirable grain for culinary purposes. It was so abundant that it flooded the market. Wild grass seeds are still harvested on a sufficient scale to reach the markets, but in former times it was done on a vast commercial scale.

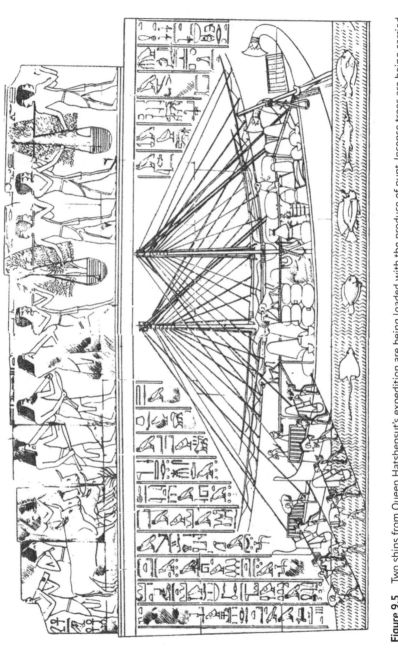

Figure 9.5 Two ships from Queen Hatshepsut's expedition are being loaded with the produce of punt. Incense trees are being carried on board in tubs large enough that four (upper register) or six (on ramp) men are required for each. This is the first government-sponsored plant introduction expedition in recorded history. From the temple of Deir El Bahari, about 1,500 BC. *Source:* Naville (1898).

In the savanna, a complex of perhaps a dozen species in mixed stands was harvested. The mixtures included species of *Panicum, Brachiaria, Eragrostis, Dactyloctenium*, and others. *Eragrostis pilosa* (L.) P. Beauv. was one of the species and it is thought to be a progenitor of *Eragrostis tef* the noble cereal of Ethiopia. Another savanna grass harvested in enormous quantities around 1900 AD was the annual African wild rice, *O. barthii* A. Chev. It was the staple of many tribes from Sudan to the Atlantic. An enormous tonnage was taken year after year from spontaneous stands. The grass was still harvested in sufficient quantities to reach the markets in the late 1900s (Harlan, personal observation). A swinging basket technique is often used and sometimes the grass is tied up in clumps before harvest, much as the Native Americans did with *Zizania* wild rice. Peoples of South America and India did the same.

In the recent past, wild grass harvests were the primary source of cereals over a large area of Saharan and sub-Saharan Africa. How important, then, is the nonshattering trait? It is certainly not necessary for harvesting in quantity, but does show up sooner or later provided seed is *planted*. Without planting, the wild-type will prevail. Rindos' (1984) idea that plants were domesticated before agriculture has no foundation in fact.

Décrue Agriculture

In French, the word for flood is *crue*; the recession of the flood when the waters go down is called *décrue*. There is no simple word in English, so the French word has been widely accepted by Africans. Farming in décrue is by no means confined to Africa, but developed its highest levels of sophistication there because the Nile, the Niger, and Sénégal rivers have very predictable and regular seasons of flooding. The Zambezi is much less regular, although décrue agriculture is practiced there to some extent. The levels of flooding on the dependable rivers vary considerably from year to year, but the timing is reasonably predictable and the farmers can plan accordingly.

Décrue farming along the Nile was relatively simple and straightforward. With the rise of the Old Kingdom, when social organization permitted public works projects, a system of levees, bunds, and dykes was established creating basins to hold flood waters and let them sink into the soil. As the waters receded, crops were sown in the moist soil and production depended on residual moisture. Rainfall was negligible. Some tomb murals show pigs being herded to trample seed into the wet soil. Others show men working the soil with a special hoe-like instrument. Vegetables and flowers were grown in special gardens with small basins laid out in checkerboard fashion and were often irrigated by men carrying pots of water (Harlan, 1986).

The most sophisticated décrue agriculture evolved in the great central delta of the Niger in Mali. There, the flood spreads out very slowly over a vast region. The slope is very slight, but the terrain is undulating, resulting in a complex hydrological system with a lacework of ponds, marshes, channels, and flat floodplains. The main region of décrue agriculture is in the Dir-Goundam-Lake Faguibine area. There, the waters may start rising in September but do not reach maximum until December–January, and it may not be until March that the waters have receded enough for an appreciable amount of planting.

In the fields farthest from the river channel, the flood comes last and leaves earliest. These are also closest to the desert dunes and are most likely to have sandy soils. The fields nearest the channel are submerged the longest and usually have clayey soils. Crops and varieties are chosen to match these conditions and vary from year to year depending on the timing and height of the flood.

The main field crops in Décrue agriculture are rice, sorghum, and pearl millet. Rice is a crop of the flood. It is sown in dry soil before the arrival of the flood or in soil moistened by summer rains. Floating varieties are planted near the river and nonfloating varieties farther away. The cultivators must make some guesses as to the height of the flood, but there are intermediate varieties that can elongate to some extent if the flood is higher than predicted.

Sorghum and pearl millet are crops of the décrue, but cowpea, roselle, cotton, and other crops may sometimes be interplanted with them. Pearl millet is the most drought resistant and is best suited to sandy soils. It is the choice for the most distant fields and the higher the flood the greater the area sown to the crop. Two major races of sorghum are used, durra for the upper sorghum fields since it is the most drought resistant and the earliest to mature, and guinea for the lower fields just above the rice zone. The sequence of crops from the dunes to the river is: short-season pearl millet, long-season pearl millet, short-season durra, longer season durra, guinea sorghum, upland type rice, intermediate rice, and floating rice.

Within this general framework, other adjustments are made. The lower fields devoted to sorghum usually do not mature before the waters rise again in the fall. To speed up the life cycle, a great deal of the sorghum is transplanted. Seedlings are grown in some sandy areas, uprooted as needed, and planted in deep holes produced by a large (ca. 1.5-m long) dibble or planting stick. The guinea sorghums used are highly tolerant of flooding and are commonly harvested from a canoe. If a flood comes early, durras and even some pearl millet may be harvested from a canoe, and most of the floating rice is gathered in this manner. The farmers might use earlier strains and avoid harvesting in the flood or damage by the "mange-mil" (*Quelea*), a small passerine bird that migrates through the region in vast cloud-like flocks in August and early September. The *time* of planting is fixed by the flood, but the strains of a crop *selected* depend on the number of days to the end of September, when most of the "mange-mil" pass through.

Agronomic practices are sensitively adjusted to height and duration of flood, soil texture, time of maturity, and so on, but there are other considerations as well. Pearl millet is preferred to sorghum as food and as much land as possible is devoted to it. In years of high flood, a great deal of pearl millet can be raised, but much less in years of low flood. Durra sorghum has relatively soft seeds and does not store well in the granary. Damage by insects and fungi is less if stored in the inflorescence, but this takes a lot of space. The people tend to eat up the durra first and save the guinea for last, since it can be stored as long as 2 years or more due to its hard vitreous seed. Naturally, it is much harder to process in the mortar, and the women do not like to pound it. The people tend to eat less of it at a meal, but it is perceived as being "strong" and more nutritious than the durra, and is used more when there is heavy work in the field. More details of the system may be found in Harlan and Pasquereau (1969).

Modification of the décrue system can be found throughout West Africa. A river is not necessarily required. Extensive areas in the savanna may stand in water during the rains. As the waters dry up, the vegetation may be burned and cereals transplanted by use of the giant dibble. Sorghum is the usual crop, and early maturing durras are the strains of choice. The plants must mature on residual moisture. Pearl millet also may be transplanted in sandy soils. In Chad and Cameroon, especially, sorghum seeds may be hoed into standing vegetation at the end of the rains, this operation being the only cultivation done for the crop.

It seems likely that the arts of décrue might have been learned in the Sahara during the "Neolithic pluvial" 6500–4500 BP. At that time, there were playa lakes that expanded in the rains and receded in the dry season, exposing land that could be farmed using residual soil moisture. Domestication of pearl millet occurred at this time in western Sahara.

Conclusions

Africa continues its reputation as difficult to know and understand, but new sequencing information is shining a light on the mystery. What we *know* is obvious enough. An impressive number of crops were taken into the domus in the African hearth. They are nearly all savanna plants, although some became adapted to the forest zone. The time range is coming into focus. Evidence from languages indicates a considerable antiquity; archaeology has given less support to the idea, but sequencing data agree. Sorghum, finger millet, and pearl millet were all domesticated 5000–4000 BP. Archaeobotanical dating of sorghum in India approximately 3000 BP are easily possible, and African crops were domesticated earlier than this. As sequencing of more crops is completed, we will know more about how long African crops have been cultivated.

There seems to be no single center of domestication in Africa; each crop is a separate story. Décrue agriculture became a finely tuned system in the central delta of the Niger, and the art of transplanting cereals to permit maturation on residual moisture was highly developed in West Africa. Wild grass seed harvesting may have begun in the Epipaleolithic, and was still an important commercial enterprise in the early 1900s. Livestock herding and pottery production were developed long before any trace of cultivated plants occurred.

References

Agrawal, D. P. (1982). *The archaeology of India.* London: Curzon Press Ltd.

Burgarella, C., Cubry, P., Kane, N. A., Varshney, R. K., Mariac, C., Liu, X., . . . Vigouroux, Y. (2018). A western Sahara centre of domestication inferred from pearl millet genomes. *Nature Ecology & Evolution, 2*, 1377–1380. https://doi.org/10.1038/s41559-018-0643-y

Butzer, W. K., Williams, M. A. J., & Faure, H. (1980). Pleistocene history of the Nile valley in Egypt and lower Nubia. In M. A. J. Williams & H. Faure (Eds.), *The Sahara and the Nile: Quaternary environments and prehistoric occupation in Northern Africa* (pp. 253–280). Rotterdam, The Netherlands: Balkema.

Clark, J. D. (1967). *Atlas of African prehistory.* Chicago, IL: University of Chicago Press.

Clark, J. D., & Brandt, S. A. (1984). *From hunters to farmers: The causes and consequences of food production in Africa.* Berkeley, CA: University of California Press.

Close, A. (1988). Current research and recent radiocarbon dates from northern Africa III. *The Journal of African History, 29*, 145–176. https://doi.org/10.1017/S0021853700023616

Conniff, K., Molden, D., Peden, D., & Awulachew, S. (2012). *The Nile River Basin: Water, agriculture, governance and livelihoods.* London: Routledge, Taylor and Francis Group. https://cgspace.cgiar.org/handle/10568/36786

Coursey, D. G. (1972). The civilizations of the yam: Interrelationships of man and yams in Africa and the Indo-Pacific region. *Archaeology & Physical Anthropology in Oceania*, *7*, 215–233.

Davis, J. M. (1987). *The archaeology of animals.* New Haven, CT: Yale University Press.

Doresse, J. (1957). *L'empire du Pretre-Jean* (Vol. *2*). Paris, France: Librairie Plan.

Ehret, C. (1979). On the antiquity of agriculture in Ethiopia. *The Journal of African History*, *20*, 161–177. https://doi.org/10.1017/S002185370001700X

Ehret, C. (1984). Historical/linguistic evidence for early African food production. In J. D. Clark & S. A. Brandt (Eds.), *From hunters to farmers: The causes and consequences of food production in Africa* (pp. 26–39). Berkeley, CA: University of California Press.

Ehret, C. (1998). *An African Classical Age: Eastern and Southern Africa in world history, 1000 B.C. to A.D. 400.* Charlottesville, VA: University Press of Virginia.

Fuller, D. Q., & Stevens, C. J. (2018). Sorghum domestication and diversification: A current archaeobotanical perspective. In A. Mercuri, A. D, Andrea, R. Fornaciari, & A. Höhn (Eds.), *Plants and people in the African past.* Amsterdam, The Netherlands: Springer. https://doi.org/10.1007/978-3-319-89839-1_19

Gowlett, J. A. J., Hedges, R. E. M., Law, I. A., & Perry, C. (1987). Radiocarbon dates from the Oxford AMS system: Datelist 5. *Archaeometry*, *29*, 125–155. https://doi.org/10.1111/j.1475-4754.1987.tb00404.x

Harlan, J. R. (1971). Agricultural origins—centers and noncenters. *Science*, *174*(4008), 468–474.

Harlan, J. R. (1986). Lettuce and the sycomore: Sex and romance in ancient Egypt. *Economic Botany*, *40*, 4–15. https://doi.org/10.1007/BF02858936

Harlan, J. R. (1989). Wild grass-seed harvesting in the Sahara and sub-Sahara of Africa. In D. R. Harris & G. C. Hillman (Eds.), *Foraging and farming: The evolution of plant exploitation* (pp. 79–98). London: Unwin Hyman.

Harlan, J. R., & Pasquereau, J. (1969). Décrue agriculture in Mali. *Economic Botany*, *23*, 70–74. https://doi.org/10.1007/BF02862973

Isaac, E. (1970). *Geography of domestication.* Englewood Cliffs, NJ: Prentice Hall.

Komroff, M. (1928). *The history of Herodotus.* New York, NY: Tudor Publishing Company.

Leblanc, M. J., Stagnitti, F., van Oevelen, P. J., Jones, C., Mofor, L. A., Razack, M., & Favreau, G. (2006). Evidence for Megalake Chad, north-central Africa, during the late Quaternary from satellite data. *Palaeogeography Palaeoclimatology Palaeoecology*, *230*, 230–242. https://doi.org/10.1016/j.palaeo.2005.07.016

Manning, K., Pelling, R., Higham, T., Schwenniger, J. T., & Fuller, D. Q. (2011). 4500-year-old domesticated pearl millet (*Pennisetum glaucum*) from the Tilemsi Valley, Mali: New insights into an alternative cereal domestication pathway. *Journal of Archaeological Science, 38*, 312–322. https://doi.org/10.1016/j.jas.2010.09.007

M'Mbogori, F. M. (2017). Farming and herding in eastern Africa: Archaeological and historical perspectives. In Oxford University Press (Ed.), *Oxford research encyclopedia of African history*. New York, NY: Oxford University Press. https://doi.org/10.1093/acrefore/9780190277734.013.134

Morrissey, A., & Scholz, C. A. (2014). Paleohydrology of Lake Turkana and its influence on the Nile River system. *Palaeogeography Palaeoclimatology Palaeoecology, 403*, 88–100. https://doi.org/10.1016/j.palaeo.2014.03.029

Munson, P. J., Harlan, J. R., deWet, J. M. J., & Stemler, A. B. J. (1976). The origins of cultivation in the South Western Sahara. In J. Harlan, J. M. J. Wet, & A. B. L. Stemler (Eds.), *Origins of African plant domestication* (pp. 187–209). The Hague, The Netherlands: Mouton.

Muzzolini, A., Aurenche, O., & Cauvin, J. (1989). La "Neolithisation" du nord de l'Afrique et ses causes. In *Neoulithisations: Proche et Moyen Orient, Méditerranée Orientale, Nord de l'Afrique, Europe méridionale, Chine, Amerique du Sud, BAR International Series 516* (pp. 145–186). Oxford, UK: B.A.R.

Naville, E. (1898). *The temple of Deir El Bahari* (Vol. 6). London: Egypt Exploration Society.

Phillipson, D. W. (1977). The excavation of Gobedra rock shelter, Axum: An early occurrence of cultivated finger millet in northern Ethiopia. *Azania, 12*, 53–82. https://doi.org/10.1080/00672707709511248

Rindos, D. (1984). *The origins of agriculture: An evolutionary perspective*. New York, NY: Academic Press.

Taffesse, A. S., Dorosh, P., & Asrat, A. (2012). Crop production in Ethiopia: Regional patterns and trends. In P. Dorosh & S. Rashid (Eds.), *Food and agriculture in Ethiopia: Progress and policy challenges*. Washington, DC: International Food Policy Research Institute. https://ideas.repec.org/h/fpr/ifpric/9780812245295-03.html

Wendorf, F., Schild, R., Hadidi, N. E., Close, A. E., Kobusiewicz, M., Wieckowska, H., . . . Haas, H. (1979). Use of barley in the Egyptian Late Paleolithic. *Science, 205*, 1341–1347. https://doi.org/10.1126/science.205.4413.1341

Winchell, F., Stevens, C. J., Murphy, C., Champion, L., & Fuller, D. Q. (2017). Evidence for Sorghum domestication in fourth millennium BC eastern Sudan: Spikelet morphology from ceramic impressions of the Butana group. *Current Anthropology, 58*, 673–683. https://doi.org/10.1086/693898

10

The Far East

According to one story the rice plant existed from the beginning, but its ears were not filled. This was the time when men lived by hunting and gathering. The goddess Kuan Yin saw that men lived in hardship and near starvation. She was moved to pity and resolved to help them. She went secretly into the rice fields and squeezed her breasts so that the milk flowed into the ears of the rice plants. Almost all of them were filled, but to complete her task she had to press so hard that a mixture

Source: A logo of the Crop Evolution Laboratory, University of Illinois; Urbana, IL.

Harlan's Crops and Man: People, Plants and Their Domestication, Third Edition.
H. Thomas Stalker, Marilyn L. Warburton, and Jack R. Harlan.
© 2021 American Society of Agronomy, Inc. and Crop Science Society of America, Inc.
Published 2021 by John Wiley & Sons, Inc.
doi:10.2135/harlancrops

of milk and blood flowed into the plants. That is why there are two kinds of rice, the white from the milk, and the red from the mixture of milk and blood.

Christie (1968)

Learning without thinking is useless. Thinking without learning is dangerous.

Confucius, sixth century BC

Archaeological Prelude

While the prehistoric record of *Homo* in China is not as long as that for Africa, it does go back to *Homo erectus*, some 1.5–1.7 million years ago, and possibly earlier still (Zhu et al., 2018). The Chinese race of that species (Peking man) differed in his tool-preparation traditions from others who ranged over Europe, Africa, and Central Asia. Peking man did not make the classic pear-shaped Acheulean "hand ax," but invented a characteristic "chopper" of his own. It looks cruder and less refined, and for a long time the Chinese *H. erectus* was thought to have been backward, but Pope (1989) suggested that the "chopper" might have been designed to cut and process bamboo, from which much more refined and elegant tools could be made. True or not, the inhabitants of China seemed to have developed a unique endemic tradition at a very early time.

Geological evidence shows that from the middle of the Pleistocene, and perhaps earlier, the hinterland of China and Central Asia had a semiarid steppe environment and occasionally may have been desiccated even further to form an arid region over vast reaches of the interior. One result was a fantastic accumulation of loess, some alluvial, but much of it windblown. There are sizeable sections of Shaanxi and Gansu provinces with loess deposits over 250 m thick and still larger areas with 150 m or more in depth. Downslope, huge tracts of Hebei, Henan, Shandong, and Anhui are covered with redeposited loess (Figure 10.1).

The end of the Pleistocene and start of the Holocene in China was, as elsewhere, a dynamic time. The Tali glacier was melting, a considerable portion of the Pleistocene fauna became extinct, sea levels rose rapidly. A climate more or less like the present emerged from 10,000 to 12,000 years ago, although sea levels continued to fluctuate by a few meters in elevation

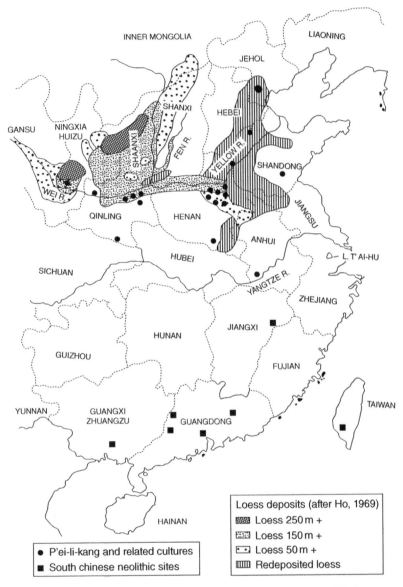

Figure 10.1 Early Neolithic sites in China. *Source:* Based on Chang (1986).

through most of the Holocene (Chang, 1986). At the end of the Pleistocene, much of China was inhabited by a variety of populations living by sophisticated hunting–gathering–fishing techniques. The different populations had different and specialized tool assemblages. The pattern was similar to that of the Near East and Africa in that it consisted of a mosaic of Mesolithic cultures showing considerable local and regional diversity, and the same trends toward smaller and more finely made stone tools.

On the loess terraces of northern China, the Mesolithic and Neolithic threshold was crossed around the middle of the seventh millennium BC. The earliest Neolithic culture of the region we now know is the P'ei-li-kang. It has been extensively studied and well dated from some 40 sites in Henan and others in Hebei and Shaanxi Provinces. The dates cluster between 8500–7000 BP. There was little change in the stone tool assemblage, and the people continued to hunt deer and other game, but by then they had domesticated pig, dog, and chicken. Crops included proso and foxtail millets and a *Brassica* species of some sort. Walnut, hazelnut, *Celtis*, acorns, and jujube were gathered and some might even have been grown. They had pottery, elaborately made grindstones, and buried the dead in special cemeteries. P'ei-li-kang culture gave rise to the well-known Yang-shao with its handsome pottery and well-developed village layouts. It spanned the 7000–5000 BP time range and flourished on the loess soils of western Henan, southern Hebei, eastern Gansu, and eastern Chinghai.

In southern China, other Neolithic cultures were evolving independently. Some caves have yielded cord-impressed pottery that may have some connection with impressed wares of Thailand and Indochina farther south. The Tseng-p'i-yen site may have the earliest pottery so far found in China. The site was dated at approximately 8400 BP by radiocarbon and between 7160 and 10,370 BP by thermoluminescence. Another south China cave, Pao-tzu-t'ou, has pottery dating to approximately 9350 BP by radiocarbon. No plant remains have been recovered, and there is little real evidence for farming. Few plant remains have been recovered, but archaeological and genetic sequencing data indicate that rice was farmed in the south of China by at least 5000 BP, before which, sago palm (*Cycas revoluta* Thunb.), bananas (*Musa* spp.) acorns, water chestnuts, and other tubers were eaten and probably cultivated (Yang, 1998).

In the Yangtze delta and around Lake T'ai-hu, yet another Neolithic culture evolved in a watery landscape. Plants grown or gathered include: water caltrop (*Trapa* spp.), lotus (*Nelumbium speciosa* Thumb), arrowroot (*Curcuma caulina* J. Graham), water "chestnut" (*Eleocharis duleis* (Burm.f.) Trin. ex Hensch), wild rice (*Zizania latifolia* (Griseb.) Hance ex F. Muell.), reed (*Phragmites* spp.), and rice (*Oryza sativa*). There are some 50 known sites in the area and the dates range from 7000–5000 BP. Much more detail is given by Chang (1986), but the general picture is that a mosaic of Neolithic cultures had evolved over most of China by 7000 BP and evident linkages began to appear by approximately 6000 BP. The Mesolithic and Neolithic threshold was crossed at least twice, once in the north and once in the south, or it could well have been crossed several times independently. The northern Neolithic was founded on the millets and the southern Neolithic was based on rice. The wetland roots and vegetables are still basic to much of Chinese cuisine. Chinese Neolithic sites appear on Taiwan dating to about 6400 BP.

The developments in northern China look very much like a center of origin and resemble the nuclear area of southwest Asia. It is a relatively small area in which early Neolithic sites are clustered and out of which a farming tradition diffused. One reason for its appearance as a nuclear area may be the loess soils that are friable, easily worked with digging sticks, are very deep, and almost totally lacking in profile so that fertility can be easily maintained. An early concentration of Neolithic sites may have been due to the concentration of loess deposits. As more research has been conducted and more information obtained, it has become clear that other Neolithic cultures were evolving elsewhere at the same time and what once looked like a small center now looks like a mosaic of contemporary developments over a wide area. Once again, the concept of a center of origin is eroding as we learn more about agricultural origins.

Recorded History

Chinese civilization, as distinct from previous cultures, can be said to have begun with the founding of the Shang dynasty sometime before 3500 BP. At about 3300 BP, the Shang capital was established at An-yang; this city was captured by the Chou tribe in 3027 BP. The dates for An-yang became important in Chinese history for several reasons. Essentially all of the Shang literature falls between 3300 and 3027 BP and consists of oracle bone inscriptions and writing on cast bronze objects. Art had reached a high state of development and Shang bronzes are world-famous for their style and technique of casting. It is from An-yang that we detect the first real evidence of contact with the West. Foreign importations of the time included wheat and barley, the horse chariot, probably the domesticated goat, and some art motifs borrowed from the Seima culture on the Volga. There is external evidence that the nomads of the Eurasian steppe were particularly active during the 13th century BC, with much warfare, raiding, and sacking of towns and cities.

Among the earliest compilations of Chinese literature is the *Book of Odes* by Shih Ching assembled from bits and fragments from the 11th century to the middle of the seventh century BC. Botanically, it is the most informative of early literatures and mentions about 150 plants as compared with 55 in Egyptian literature, 83 in the Bible, and 63 in Homer (Ho, 1969). In the *Book of Odes, Panicum* millet is mentioned 27 times, the mulberry 20 times, and *Artemisia* is mentioned 19 times with some 10 varieties.

The soybean is first mentioned in 2664 BP in connection with tribute paid to the Chou by the Shan-Jung (Mountain Jung) tribe. Hemp (*Cannabis*) was not mentioned in the Shang oracle literature, but occurs seven times in the *Book of Odes*. Iron implements for agriculture became significant about 2400 BP. Manuring, crop rotation, double cropping, and intensive agriculture as well as the first large-scale irrigation projects all date to the third century BC (Ho, 1974).

After Alexander the Great (died 2323 BP) and the establishment of Greek states from Afghanistan to the Mediterranean, regular contact was maintained between China and Persia by way of the silk routes. Laufer, in his scholarly study Sino-Iranica (1919), traces the arrival in China of a number of Near East cultigens: alfalfa and grape were introduced in 2126 BP; cucumber, pea, spinach, broad bean, chive, coriander, fig, safflower, sesame, and pomegranate arrived from Iran at various times from the second to the seventh centuries AD.

The Chinese crops were very slow to spread out from their homeland. The millets constitute a special case that will be discussed later, but cultigens of certain Chinese origin were unknown to the West until very late. The peach is said to have reached India by about the second century AD. Many authors credit the Chinese with domestication of the apricot, but since the wild races range from Turkey to China it seems likely that other people were also involved.

The West did not know rice until after the era of Alexander the Great. Theophrastus (as translated by Arthur Hort, 1916) gave a good description of it and called it the emmer of the Indians. Overall, Far Eastern agriculture may be characterized as introverted with very little dispersal until well into modern historical times, and many crops did not move out until the arrival of European shipping in the late 15th and early 16th centuries AD. There was, in fact, a notable lack of long-range diffusion, as illustrated by the fact that the Chinese were casting iron for 2000 years and using the crossbow for 1000 years before the Europeans began to use them.

Far Eastern Crops

If we look at the crops individually and in some detail, they present a non-centric pattern rather like that in Africa. Plants were domesticated out of the native flora wherever the people found them. For that reason they may be grouped according to ecological adaptation, which surely reflects something about their origins.

Northern China

The agriculture that evolved on the north China uplands was based on the millets, soybean, and a suite of fruits and vegetables. Li (1970) pointed out that several of the ancient vegetables are no longer grown, but linger on as weeds of waste places or in fields of modern crops. The most important vegetable to the ancient Chinese was a mallow, *Malva verticillata* L. Others that were once cultivated, then abandoned include *Angelica japonica* A. Gray., *Crepidiastrum denticulatum* (Houtt.) Pak and Kawano, *Rorippa indica* (L.) Hiern, *Persicaria hydropiper* (L.) Delarbre., *Viola verecunda* A. Gray, and *Xanthium strumarium* L. (Li, 1970). Modern vegetables have many species represented by the Cruciferae (*Brassica*, radishes, etc.) and Allium (onion, leek, chive, shallot, garlic, etc.). They also include a number imported from abroad, such as pea, cowpea, lettuce, and so on, but the imports have been modified in typical Chinese style. Cowpeas are used as green pods, rather than dry seed, and are typically of the "yardlong" types. Peas are also used in the green pod form, and the famous Chinese snowpea is the result. Lettuce was selected for succulent nonbitter stems and reduced leaves, and is very different in appearance from western lettuce.

The fruits of north China were selected from the temperate forest flora and many belong to the Rosaceae. Apples, pears, plums, cherries, peach, apricot, and hawthorn, all in great diversity, make up most of the inventory. To this can be added persimmon (*Diospyros kaki* Thunb.) and the jujube (*Ziziphus jujuba* Mill.). Mulberries were and still are grown more for silkworm fodder than for their fruits.

Cannabis (hemp) was a north China domesticate, but could well have been taken into the domus elsewhere. It was the principal coarse fiber of the region; the grains could be eaten and oil can be expressed from the seeds as well. Its narcotic properties were also known. Another technical plant was *Toxicodendron vernicifluum* (Stokes) F. A. Barkley (syn. *Rhus verniciflua*), the lac plant. Silk production and weaving and lacquerware are two uniquely Chinese artistic crafts that have contributed much to the elegance of the high civilization that emerged in north China.

The millets and soybean are covered in later sections dealing with specific crops.

Eastern China Coastal Plain

In the watery lowlands of the east China coastal plain, another group of plants was selected from the native flora for domestication. From the

Nymphaceae came at least three: the oriental lotus (*Nelumbo nucifera* Gaertn.), the prickly water lily (*Euryale ferox* Salisb.), and the water shield (*Brasenia schreberi* J.F. Gmel.). Others include the popular water "chestnut", the water caltrop, a water mustard [*Brassica rapa* L. subsp. *nipposinica* (L. H. Bailey) Hanelt], *Oenanthe javanica* (Blume) DC, arrowroot, *Ipomoea aquatica* Forssk. (a sort of morning glory with edible leaves), wild rice, and common rice. The rhizomes of reed and cattail were and are still gathered as well.

Wild rice belongs to the same genus as the American wild rice and it is said that it was once grown as a cereal in north China. It became infected with a smut (*Ustilago*) that causes the stems to swell and results in sterility. The plant is now grown as a vegetable and is propagated vegetatively. As we have seen, the wetland suite of plants was domesticated in the Neolithic or earlier time and after some thousands of years is still a major part of Chinese culture.

Southern China

In the south of China, still more plants were domesticated. These include several species of *Brassica*, the red bean [*Vigna angularis* (Willd.) Ohwi & H. Ohashi], velvet bean (*Mucuna pruriens* (L.) DC. var. utilis (Wall. ex Wight) Baker ex Burck), the Chinese yam [*Dioscorea esculenta* (Lour.) Burkill], the day lily, and others. However, the major contributions from the south came from the subtropical forest and woodlands. Sour orange, sweet orange, mandarin orange, kumquat, loquat, wampi, litchi, *Canarium*, and other fruits were brought into the fold. Fiber plants included *Boehmeria nivea* (L.) Gaudich, *Abutilon theophrasti* Medik., and *Pueraria lobate* (Willdenow) Ohwi. The tung trees (*Aleurites* spp.) produce commercial oil, and tea is a major crop on the world scene.

Asia and South Pacific

As we move into the tropics south of China, we find contributions coming from different ecological zones, as expected. The savannas with their long dry seasons present us with annual cereals like rice and *Coix* and with yams that behave like annuals. Wetlands yield aroids like *Colocasia* and *Alocasia*. The forest margins provide cucurbits such as *Mamordica*, *Benincasa*, *Trichosanthes*, *Luffa*, and the ubiquitous bottle gourd. More species of *Citrus* are added. Cinnamon, ginger, turmeric, black pepper, and others add spice to food, and the betel nut–betel leaf combination becomes a popular masticatory.

The Malay Peninsula and the South Pacific Islands have provided a bewildering array of fruits, roots, and spices. Some of the best known and most popular include: mango, mangosteen, durian, rambutan, jambos, sugarcane, nutmeg, clove, and coconut. The coconut is primarily a coastal plant and probably of island origin, the rest tend to be forest margin plants adapted to more light and less shading than is found in the rainforest climax.

Man in Southeast Asia and Indonesia left a record of respectable antiquity as indicated by *H. erectus* remains in Java (Pithecanthropus or Java man), but much later sites are of more interest with respect to agricultural origins. At the end of the Pleistocene to start of the Holocene juncture, the land configuration of Southeast Asia and adjacent islands was very different from the present. Islands on the Sunda shelf, west of the Wallace Line were joined to the mainland. Sahul (the island continent of New Guinea, Australia, and Tasmania) formed a separate land mass. Sahul had been inhabited by anatomically modern man since 40,000 years ago, or possibly earlier. Even at glacial maximum, with sea levels approximately 90 m below present, there were water gaps of some 100 km or more between Sunda (Malay Peninsula and the islands of Borneo, Java, and Sumatra) and Sahul. Somehow, Pleistocene man was able to cross the gaps and become a representative of a third order of placental mammals to inhabit Sahul—after bats and rats. This early colonization extended to New Ireland (an island of Papua New Guinea) with a date of approximately 32,000 years ago and to the Solomon Islands by 29,000 years ago (Allen et al., 1989). The Pleistocene seafarers will come up again later.

No doubt, people were inhabiting margins of the ancient shorelines in the early Holocene, but nearly all traces have been drowned by rising seas. Here and there, coastal uplifting has exposed Pleistocene shorelines, but, in general, the archaeological record of early coastal settlement has been buried by sea water. Tasmania became separated from Australia about 12,000 years ago, and New Guinea from Australia about 8000 years ago. More or less present sea levels were reached approximately 6000 years ago (White & Connell, 1982). The early Holocene evidence is, therefore, closed to us although a fair number of later sites are known and have been studied (Higham & Maloney, 1989).

In the uplands on the mainland, evidence of exploitation of local plant and animal resources has been uncovered. Of special interest is the Hoabin-hian culture, named by Madeleine Calani in 1927 after the province of Hoa Binh in old Tonkin. A number of sites have been located in Myanmar (Burma), Thailand, and Vietnam. Among the earliest is Spirit Cave dating to over 11,000 BP and other sites like Banyan Valley Cave lingering on as late as 900 AD (Higham, 1976; Hutterer, 1983). Among the plant remains found in Spirit Cave were seeds or fruit fragments of *Aleurites, Canarium, Madhuca, Terminalia, Castanopsis, Cucumis, Lagenaria, Trapa, Areca,* and *Piper*. A few other plants were reported in the original report that seemed ecologically and geographically out of place and were challenged, but the list of tropical materials is impressive. A second excavation added *Celtis, Ricinus, Mamordica, Nelumbium,* and *Trichosanihes* or *Luffa*. The Ban Kao caves in Thailand yielded remains of *Licuala* (a palm), and *Syzygium* (an edible fruit), the bark of which has medicinal properties (Pyramarn, 1989). The Hoabinhians preferred karst topography, hunted a variety of animals, and were noted for a pebble tool and large flake technology. Of course, they may also have been making sophisticated tools, traps, snares, and nets of bamboo, woody vines, and other perishable materials. The economy is described as broad-spectrum hunting–gathering and their late persistence among farming people is of interest; there is no good evidence that agriculture was practiced there (Yi et al., 2008).

Glover's work on Timor and Sulawesi reported that dry caves were occupied from 14,000 BP to about 0 AD (Glover, 1986). Before 5000 BP, plant remains included *Aleurites, Celtis, Areca, Coix,* and *Piper* and after 5000 BP, no *Celtis* but *Inoearpus,* bamboo, *Lagenaria,* and possibly *Setaria* (Hutterer, 1983). On the west side of the region, the cave site of Beli-Lena in Sri Lanka yielded seeds of wild breadfruit and wild bananas, dating to 10,000–12,000 BP.

Thus, we are developing evidence for people in the early to mid-Holocene living in caves, hunting, and exploiting tropical forest products. None of the sites has yielded any real evidence for early agriculture. Sites that do reveal unequivocal evidence of farming are relatively late. At Non Nok Tha and Ban Chiang in Thailand, the people had domestic cattle, pig, dog, water buffalo, rice, and cord-impressed pottery. Dating has had some problems, but the sites were probably of the fourth millennium BC. Xom Tria cave in northwest Vietnam yielded domesticated rice dated to about 5000 BP (Chang, 1989).

Other agricultural sites are later, and probably occurred only after introduction of agriculture from populations from southern China (Bellimod, 2004). Location of sites mentioned are shown in Figure 10.2.

Meanwhile, something was going on in New Guinea. There is some evidence for an early attack on the tropical forest. Heavy axe heads, "waisted" and ground, make an appearance by 26,000 years ago, and pollen sequences show disturbance of the forest of some kind by 30,000 years ago (Groube, 1989). What this means in terms of forest exploitation, we do not

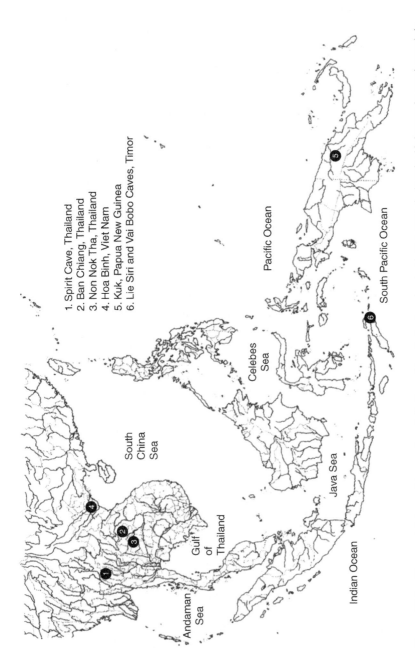

1. Spirit Cave, Thailand
2. Ban Chiang, Thailand
3. Non Nok Tha, Thailand
4. Hoa Binh, Viet Nam
5. Kuk, Papua New Guinea
6. Lie Siri and Vai Bobo Caves, Timor

South China Sea

Pacific Ocean

Andaman Sea

Gulf of Thailand

Celebes Sea

Java Sea

Indian Ocean

South Pacific Ocean

Figure 10.2 Sites containing evidence of exploitation of local plant and animal resources during the Holocene in southeast Asia and the Southern Pacific areas.

know, and the disturbance could be by natural causes. Certainly, other tropical rain forests were altered by ice-age change of climate during the Pleistocene. At Kuk Swamp in the highlands, clear evidence of landform modification turns up dating to about 9000 BP.

Buried under peat in a natural swamp in New Guinea, Golson (1984) found an extensive drainage and irrigation canal system. One canal was some 10 km long, 1 m deep, and 10 m wide. There were extensive fields with raised beds surrounded by ditches for water control either for drainage or irrigation. One immediately thinks of taro, *Colocasia*, or *Cyrtosperma* production, but no plant remains have been found. Ample evidence for the presence of pigs appeared by 6000 BP. The pig, as a placental mammal, was absent from New Guinea and was almost certainly brought by man before that date. It was taken everywhere in the Pacific areas that were colonized by farmers. The source is obscure because the date seems to be earlier than farming is attested to in the Philippines and Borneo (Bellwood, 1985). It has been suggested by Denham et al. (2004) that agriculture in Papua New Guinea began at least 6950–6440 BP and was not imported from Southeast Asia, but emerged independently in the Highlands. At about 6000 BP, the rather Hoabinhian-like core-and-large-flake industry of Australia was suddenly changed to a highly developed microlithic complex, but if this was introduced, then the pig was left behind (Moore, 1976).

The implications of the irrigation and drainage systems are clearly in favor of some independent experimentation with horticulture. The effort required to put in the systems using digging sticks was considerable. There must have been some worthwhile reward. Colonies of spontaneous taro on the island have long been considered to be recent escapes, but they prove to be diploids, while the common cultigen of Southeast Asia and India is a triploid (Jones & Meehan, 1989). The case is not proven, of course, but it is a strong one for another independent origin of plant husbandry if not domestication.

Some years ago, Hutterer (1976) pointed out that the archaeological evidence in Southeast Asia was anomalous in both time and space: ". . . the basic problem is the lack of uniformity of cultural development in Southeast Asia—the apparent impossibility of identifying regional traditions and local sub traditions within a generally valid chronological framework." He then went on to suggest this might be due to the extreme difficulty of living in a rain forest yearlong without supplementation from agriculture. The usual shortage in this environment is in carbohydrates, and most hunter-gatherers of the rain forest trade with farmers for the supplements. Various local cultures work out their own solutions, resulting in a very complex mosaic pattern lacking in evident integration.

Of course, there were other environments in Southeast Asia. There were savannas and forest-savanna ecotones that offer far more food resources than the rain forest, and it is from such environments that most of the indigenous domesticates came.

The archaeological evidence to date does not document a transition from hunting–gathering to farming in Southeast Asia nor is there sure evidence of an indigenous development. This may be partly due to insufficient excavation, but could also be due to the manner in which the threshold into the Neolithic was passed. The Hoabinhian people kept their traditions long after their neighbors were farming and there was, no doubt, considerable interaction among the different cultures. The spread of farming systems across the Pacific islands has been reviewed several times (Bellwood, 1985; Denham et al., 2004; Kirch, 1982; Spriggs, 1989). The picture may change in the future with more information.

The oldest Neolithic on Taiwan is of the Ta-p'en-k'eng tradition from southeastern China. Cord-marked pottery was present, which may have some relation to the cord-impressed wares of Indochina. The tradition on the island began about 6400 BP and continued to about 3500 BP. Early Neolithic, as of current information, is no earlier than 5500 BP in the Philippines and probably later (Balbaligo, 2015). It had definitely been established on Luzon by 5000 BP; on Mindanao, Borneo, Sulawesi, and Maluku by 5500 BP; and on Timor soon after (Bellwood, 1985; Spriggs, 1989). The agricultural complex at this time included pig, chicken, breadfruit, *Alocasia*, taro, yams, bananas, sago, and betel nut. Cereal culture declined as farmers moved eastward. The millets dropped out one by one, Coix alone reaching New Guinea, and rice stopping in the main Indonesian islands. New Guinea had its own Neolithic and was skirted by the Polynesians. By this time sailing canoes were available and long distance travel was possible.

The peopling of islands beyond New Guinea and the immediate islands (New Ireland, New Britain, etc.) by farmers began with the Lapita complex, approximately 3600–3500 BP. The Bismark Archipelago, Santa Cruz Island, New Hebrides, New Caledonia, Fiji, Tonga, and Samoa were settled by Lapita horticulturalists. They had a distinctive pottery and used characteristic stone adzes for woodworking. Pig, dog, and fowl were domesticated and plants included taro, yams, breadfruit, bananas, and plantains and a varying selection of other food crops.

The Marquesas may have been reached by 2150 BP, and eastern Polynesia no later than 300 AD. The earliest ^{14}C dates are: Marquesas 2100 BP; Society Islands 900 and 890 AD; Hawaii 390, 610, 795 AD; New Zealand 1050 and 1230 AD; Cook Island 1020 AD; and Easter Island by 400 AD. Pottery was abandoned at the outset of this expansion (Kirch, 1982). In general, the beginnings of agriculture in islands of Southeast Asia is a complicated story, and linguistic and archaeological data suggest that more than one route and process blended to bring the cultivation of crops to the area (Barker & Richards, 2013).

The Millets

Setaria italica and *Panicum miliaceum* were basic to the north Chinese Neolithic, and were presumably domesticated in the region, yet both are found in a sprinkling of Neolithic village sites throughout Europe through the fourth millennium BC. They were seldom important components of plant remains. They were reported only from Niederwil among the Swiss Lake dwellers, for example; but they occurred in enough sites that there can be little doubt of the presence of both millets during the fourth millennium BC in Europe. *Panicum* has also been reported from Jemdet Nasr, Mesopotamia (5000 BP) and possibly Argissa-Maghula, Greece at about 7500 BP (Renfrew, 1969).

The only European culture that grew *Panicum miliaceum* extensively was the Tripolye of the Ukraine. The culture flourished from 5,800 to 4,900 BP and *Panicum* was one of the major crops. Neither millet has been studied intensively, and the archaeological studies of the vast Eurasian steppe between China and the Ukraine are not yet sufficiently advanced for us to choose between possible alternatives. The possibilities are: (a) the millets were domesticated in China and dispersed to Europe before 6,000 BP in Neolithic times, (b) they were domesticated in the West and were dispersed to China, and (c) there was more than one domestication. The presumptive progenitor of *Setaria italica* is *S. viridis* (L.) P. Beauv., a ubiquitous weed from Japan to England and now widespread in North America and elsewhere. It is frequently stated that the progenitor of *P. miliaceum* is not known, but in the Flora of USSR (Botanical Institute of the Soviet Academy of Sciences, 1968–1973) it is reported as weedy, naturalized, or escaped, and common from European Russia to eastern Siberia (see also Komarov, 1934). Without very careful analysis, it is often difficult to separate wild from weedy races.

Both millets are adapted to the summer rainfall belt of temperate Eurasia. They were well known to the Greeks and Romans and to Indians of ancient times. It may be noted, however, that a number of Indian names for panicum suggest that it came to India from China. In Sanskrit the name is cinaka (meaning Chinese); Hindi, chena, cheen; Bengali, cheena; and Gujarati, chino. The Persian word is essentially the same as the Chinese, shu-shu (Laufer, 1919).

No other known crops had such a distribution in that time range. Wide dispersals in the fifth millennium BC are certainly possible, but one might have expected more than the two millets if this was the explanation. The rather slow spread of agriculture across Europe at that time suggests that early European farmers were having enough trouble just getting across Europe without attempting the much longer trip to China. The Chinese crops, as we have seen, did not disperse much until very late.

In our present state of ignorance, independent domestications appear to be the most likely answer, but new information could easily lead to other conclusions.

Soybean

The wild soybean is a small, slender creeping vine bearing a few small pods with small, black seeds. The plant is widely distributed from southern Siberia, through Manchuria, throughout the eastern coastal plain of China, and westward to Szechuan. It is rather weedy, and is often found in city parks under the shade of trees. Presumably, it was once a woodland or temperate forest plant before the natural vegetation was removed for agriculture. The changes under domestication have been enormous.

Evolution of bush types from vines is common under domestication. It has happened in American beans, African cowpeas, peanuts, oriental soybeans, and others. The viny ancestral types of soybean are usually retained as well, and trailing forms are still grown for fodder. Soybeans are prepared for food in many ways, often including fermentation with special strains of yeasts and fungi cultured for the purpose. Tofu and soy sauce are the products best known outside the Orient. Soybean is a major Chinese contributor to agriculture, and is widely consumed in China; however, they produce only about 15 million tons (2018) and import more that 90 million tons. Soybean production, which puts the crop among the top 10 in the world, comes from the United States and Brazil where most of this crop is exported or fed to animals.

Rice

There were two cereals in the ancient agricultural system of Southeast Asia, rice and *Coix* (Job's tears). There is some evidence that *Coix* was the older of the two, at least in the rain forest zones. It spread into regions of the Philippines, Borneo, and New Guinea where rice did not reach. The great swampy deltas of Southeast Asia probably supported vast stands of wild rice at one time, but the rice in such environments was mostly, if not exclusively, of the perennial floating kinds. Seed was no doubt harvested by gatherers using canoes or from the tangled mass after flood waters receded at the end of the rains. The perennial races, however, are poor seeders compared to the annuals, and the environment is extremely difficult to exploit for agriculture. It is most unlikely that rice was domesticated in the delta zones, which must have been sparsely settled until social and political systems evolved permitting the construction of dikes, canals, and other water control measures. Rice in Africa was domesticated in the savanna; the situation in Asia must have been similar.

The domestication and diffusion of Asian rice has been reviewed several times by T. T. Chang, and Figure 10.3 is adapted from one of his papers and graphically shows the most likely region of the origin and subsequent dispersal (Chang, 1989). H. I. Oka and more recently McCouch and colleagues studied the genetics of rice domestication for many years and anyone interested should consult Kovach et al. (2007), Oka (1988), and Rawal et al. (2018).

Within modern historical times the traditional digging stick and hoe shifting cultivation has been replaced in many places by wet rice cultivation. The lower swamp areas in particular were intensively cultivated only in the last century or two and much of the settlement was so late that some census figures are available. In the 1850s, Burma planted about 607,500 ha of rice. The figure topped out at over 4.86 million ha at the beginning of the 21st century, but has decreased since to around 2.63 million ha as of 2016, most of it in the lower Burma Irrawadi swampland delta area. Yield per acre has nearly tripled in that time (FAO statistics database, http://www.fao.org/faostat/en/#home). Thailand in the 1880s reported about 1 million ha of rice, and today about 12 million are planted. Much of the increase has been in the lower Chao Phraya Valley (the Bangkok plain). In Vietnam, diking and draining of the Mekong delta began in the 19th century. The delta as a whole is still not fully occupied but the increase in rice acreage has been phenomenal. Today in Vietnam, about 6.48 million ha, out of 8 million ha in cultivation, are planted with rice. The rice areas of Indonesia total about 14 million ha, much of it in Java and in the mangrove swampland of northeastern Sumatra. In the Philippines, there are about 5.6 million ha of production in both shifting cultivation and wet rice land. In Laos, much of the agriculture was still of the original indigenous kind until the 1980s. Most of the country is mountainous upland, but rice production has increased to just shy of 1 million ha. Malaysia's population has increased very rapidly and does not permit the development of plantation crops on a large scale, and they grow only ~283,000 ha of rice, leading to a necessity of importing nearly a million tons in 2018 (https://www.fas.usda.gov/data/

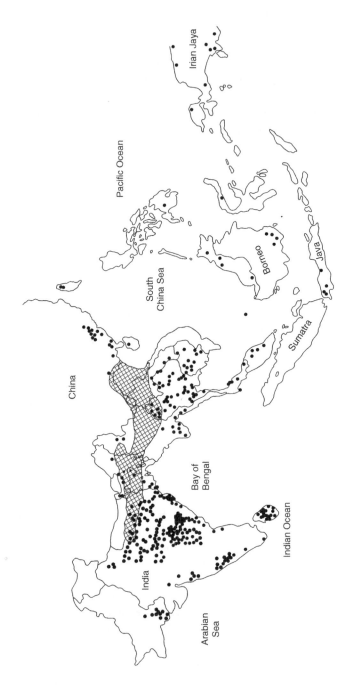

Figure 10.3 Known distribution of wild and domesticated rice in Asia. The gaps in Burma and Vietnam are probably due to inadequate collection (Harlan, 1975). The cross-hatched area marks the center of diversity of cultivars. *Source:* Adapted from Chang (1989).

malaysia-grain-and-feed-annual-1). On the whole, wetland rice dominates landscapes and has replaced shifting cultivation of root and tree crops in Southeast Asia in recent centuries.

Although the origin and antiquity of wetland rice cultural techniques are uncertain, there seems to be little question that wet-field taro (*Colocasia* spp.) production had been developed earlier. Well-engineered, but small-scale terraces were constructed and water was led by canals to flood them. Such practices persist among tribes in New Guinea who have not yet taken up rice growing. The suggestion has been made that rice was domesticated from wild or weed races that infested the flooded taro fields. This has merit for limited regions of the Southeast Asia mainland and the larger islands, but if this sequence took place it was probably in addition to rice domestication in the savanna zones with prolonged dry seasons as in India and southeastern China, Burma, and Thailand.

The evidence for early rice cultivation in Thailand suggests that the cleavage between cereal agriculture and vegeculture is exaggerated and may never have been real. The fact is that there are more tropical cereals than temperate ones and that cereals play at least some role in most tropical agricultural systems. Wild rice is a food resource that could not be overlooked by gathering people and, as we have seen, is still harvested. Rice was domesticated 8000–12,000 years ago (Molina et al., 2011). The archaeological evidence suggests that domesticated rice was widespread in China, India, and Southeast Asia by 7000 BP. Rice had been available for a long time, but rice-based agriculture did not become really expansive until population densities and social and political structures were such that intensive agriculture was not only demanded but could be practiced. The historical evidence of Ho (1974) indicates intensive agriculture got under way in China in the third century BC; when it developed in Southeast Asia we do not know.

Sugarcane

The basis of modern sugarcane production probably started in New Guinea as a mutation in *Saccharum robustum* E. W. Brandes and Jeswiet ex Grassl. This species is an octoploid with 80 chromosomes and has large, stout, hard canes. It is often used to build fences in New Guinea to keep pigs out of the garden or to keep them in the pen. At some time, mutations occurred that blocked the pathway from sugar to starch, and sweet canes resulted. These were the "noble" canes that were taken across the Pacific Ocean by Polynesians and then diffused westward to Southeast Asia. In southern China, and perhaps in Assam, the cultivated canes came into contact with wild *Saccharum spontaneum*, a widespread weedy plant of Asia and Africa. As a result of introgression, the thin cane types evolved. These have been grouped under the epithet *Saccharum sinense* Roxb. and have variable chromosome numbers from $2n = 82 - 124$ or thereabouts (Simmonds, 1976b). The noble canes were traditionally exploited by chewing, and syrup was extracted from the thin canes. The syrup could also be boiled down to make a dark brown sugar. Arabs brought sugarcane to southern Spain after their invasion of Europe, and the Spaniards introduced it to the New World, where sugar production became intimately associated with slavery. The abundant, cheap, white, refined sugar is a recent development on the world scene, and sugarcane has become one of the world's largest crops by production quantity.

Bananas and Plantains

The wild diploid bananas are native to the Malay Peninsula and major Pacific islands, especially Borneo. They reproduce by seeds which are numerous and extremely hard, rendering the fruits unattractive as food. Some genotypes, however, appeared that are parthenocarpic, that is, produced fruits without seeds. They were, of course, sterile and had to be vegetatively propagated and would have died out without human assistance. There are two species involved in the evolution of commercial bananas, *Musa acuminata* Colla (A genome) and *M. balbisiana* Colla (B genome). The A genome tends to produce the sweet table or dessert banana; the B genome tends to produce the dry cooking banana or plantain. Natural hybridization and polyploidy have produced the combinations: AB, AAB, ABB, AAA, AABB, and ABBB. The most common genomes in production are AA, AAA, AAB, and ABB; the others being relatively rare. AAAA is not known to have arisen naturally although it is frequently produced in breeding programs (Simmonds, 1976a).

We do not know when bananas were first cultivated. They spread from their homeland to India and southern China, probably rather early. Polynesians took them to Madagascar when they colonized the island at the beginning of the Christian era. Polynesians also took it to remote islands of the Pacific. They made their way to Uganda where an important banana-based culture evolved and they were found on the west coast of Africa when Portuguese began to explore it in the 15th century. The Arabs knew of it and called it muza, hence *Musa* and Musaceae.

Bananas of the Australomusa group are native to New Guinea and adjacent islands and were taken by Polynesians to some of the islands they colonized. They are seedy and some have escaped and become naturalized in Tahiti, for example. Another species of the Australomusa group is *Musa textilis* Née, or Manila hemp. This was once a major source of marine cordage. The long, strong fibers are resistant to sea water and are extracted from the leaf sheaths. Still another member of the family is important in Ethiopia. *Ensete ventricosum* (Welw.) Cheesman (Ethiopian banana) was domesticated and became a staple of some Ethiopian tribes. The seedy fruits of enset are not eaten, but the starchy stem base can be processed into flour.

Coconut

There has been some dispute and controversy over the origin of the coconut, some of it due to poor taxonomy and much due to lack of information. One theory is that there have been several independent origins of coconut cultivation in the Pacific region and in the Indo-Atlantic regions. Gunn et al. (2011) suggested that there are two geographic origins of coconut cultivation including the South Pacific Islands from New Guinea westward through Sulawesi, Borneo, and northward through the Philippines and a second origin in southern India. The large fruits with thick fibrous husks and hard shell are adapted for dispersal by sea. Violent storms have probably played a role in casting the nuts far beyond normal high tide where they may sprout in great numbers. Man has also played a major role in both dispersal and in the selection of special types. The exploitation of the coconut as a plantation crop is a recent industrial age phenomenon. In traditional agriculture, it was fundamental for the settling of the smaller Pacific islands and atolls. The Pacific islands had few natural resources and Polynesian colonizers had to establish their cultigens before many of them could be settled. Coconut, pandanus, *Cyrtosperma*, and taro (*Colocasia esculenta* (L.) Schott.) were among the most important on low, relatively dry islands where bananas, breadfruit, and sugarcane did not do well. Excavation to reach fresh water lenses was often required for cultivation of the tubers.

The coconut was reported on Cocos (or Keeling) Islands and the west coast of Panama by the Spanish natural historian Gonzalo Fernandez de Oviedo y Valdes (1944 edition). His first description was written in 1519, giving almost no time for a European introduction of the plant. The possibilities are: (a) Ovideo's account was garbled and the coconut was not there, (b) the coconut reached the west coast of Central America through natural means of dispersal not very long before Columbus, and (c) the plant was brought to the area by human agency not very long before Columbus.

Oviedo's account was, indeed, garbled and early European observers gave contradictory accounts. Oviedo's description of the coconut is accurate, detailed, and unmistakable, but he then supplies us with a line drawing of a palm which is not a coconut. He said it was particularly abundant in Cacique

Chiman on the west coast of Panama, yet Wafer (1934 edition), a very good and reliable observer, passed through Cacique Chiman in the 1680s and could not remember seeing a coconut on the mainland. Wafer did, however, report a very strange account of a "frolik" on Cocos Islands, by several of the ship's company in which they cut down a number of "coconut" trees, harvested about 80 L of milk, and drank until they were benumbed. Cook (1939) found a palm on Cocos Islands that is "remarkably similar in size and appearance" to the coconut, but is entirely unrelated. Despite the garbled reports, it seems most likely that the coconut had, in fact, established a foothold on American shores before the arrival of Europeans. Johnathan Sauer (personal communication) studied the populations in Panama and considers them to be naturally introduced.

Orange

Little is known of the early history of the orange. It is thought to have originated in northeastern India and western Burma. Its adaptation to the subtropics with some frost resistance suggests that it came from the hill country, and perhaps its range extended into southern China. The sweet orange (*Citrus* × *sinensis* L. Osbeck) probably arose as a mutant of the sour orange (also known as *Citrus* × *aurantium*) that originated through an F_1 cross between two other citrus species, a pummelo and a wild mandarin (Wu et al., 2014). The latter is rather weedy and readily distributed by parrots in both the Old and New Worlds. It was introduced to southern Europe in late historical times, but introduced to the New World by 1500. Seed of most oranges are produced asexually, but there is sufficient sexuality in the species for breeding programs. As a fruit of commerce, it is a recent development, but we know little of its antiquity.

The original name began with an "n"; late Sanskrit, naranga; Hindi, narangi; Persian, narang; Arabic, naranj; and so on. The "n" is preserved in Spanish, naranja, but lost in French, orange, Italian arancia, and in English due to the absorption of the "n" of indefinite articles: an-norange, une-orange, and so on. The House of Orange, the royal line of Holland, derived its name from a town on the Rhone River in France. In some regions, regardless of the local language spoken, the orange is called portugal, because it is perceived as being introduced by the Portuguese.

Mango

The genus *Mangifera* is native to the Far East region and the wild forms confined to it. The greatest number of species is found on the Malay

Peninsula, Borneo, and Sumatra. *Mangifera indica*, however, is found wild primarily in northeastern India and into western Burma. Domestication seems to have been straightforward without contributions from other species. The wild forms are highly fibrous and have a strong taste of resin or "turpentine". Selection has been toward less fiber, more juiciness, and less resin. Some cultivars are exceptionally fine and their fruit enjoys great popularity throughout the tropics. Distribution out of India into Southeast Asia and offshore islands may have been in first millennium BC, but dispersal to the rest of the world has been only in recent centuries.

Yams

Several species of yam, *Dioscorea*, are cultivated in the Far East but by far the most important are *D. esculenta* (Lour.) Burkill and *D. alata* L. They have "annual" type tubers and derive from savanna environments with a long dry season. Both species have complex polyploid series with a base of $x = 10$, while the African domesticates have a base of $x = 9$. The principal Asian yams probably originated in the north-central part of the Malay Peninsula. Papua New Guinea is probably the center of diversity today, but that is, in part, due to a decline in importance of yams on the mainland.

In the Pacific Island area, yam is considered a dry crop and taro a wet crop. Soil is mounded or ridged up for yams; ditches are dug to keep taro wet. Males tend yams and females are excluded from yam gardens. Taro is cultivated by females only; males, even babes-in-arms, are excluded from taro gardens. There are sexual implications in the phallic shape of the yam tubers and the vaginal shape of the taro leaf. In New Guinea, ceremonial yams are grown each year with great care and ceremony. Contests are held among villages to see who can grow the longest and heaviest tuber and records are kept from year to year. The male gardeners who tend the ceremonial yams must refrain from sexual intercourse while the yams are growing and other rites are practiced. These and other details are discussed in a charming essay by Barrau (1965); see also Coursey (1976).

Hunter-Gatherers of Japan

Japan was hardly at the cutting edge of plant domestication, but provides a fine illustration of an advanced lifestyle of a hunting–gathering people. The Jomon period goes back to some 12,000 years BP. The culture evolved over time, but near the beginning produced some of the earliest pottery yet recorded anywhere on earth. They not only had early pottery; they had pottery with a flair and of great artistic merit. They lived in villages and developed a very dense population for hunter-gatherers. They evidently were exploiting their resources rather fully as shown by a sudden increase in population after they discovered how to detoxify horse chestnut (*Aeschulus*). Piles of horse chestnut shells began to appear in sites about mid-Jomon or approximately 7000 BP. The plant is still exploited in the wild, and there is a close correlation between regions where this is done today and regions where archaeology shows it was done thousands of years ago.

The culture was evidently well adapted to the temperate forest resources of the islands, as evidenced by its long history. There are many parallels between the Jomon and the affluent hunter-gatherers of the northwest coasts of North America (Koyama & Thomas, 1982). The way of life seems to have been very satisfactory and there was probably no compelling reason to take up agriculture. About 2300 BP, however, a Yayoi culture arrived from the mainland with a rice agriculture, together with other crops, and swept over the islands in a few centuries.

Plant Domestication in India

In the plan of presentation in this book, India falls between the Near East and the Far East. It seems also to have that position in terms of the evolution of agriculture. Wheat, barley, lentil, chickpea, *Lathyrus*, flax, and other crops, arrived from the west and the Harappan civilization arose. Rice, fruits, and roots arrived from the east. Sorghum, pearl millet, finger millet, and cowpeas came from Africa. India was receptive to all; the north Gangetic Plain adopted the Near Eastern complex, and peninsular India accepted the tropical complex of Southeast Asia. However, the Indians also domesticated plants from the local flora and archaeological evidence suggests the earliest agriculture in India was based off domesticated regional plants, with later incorporations of crops from other regions between 5000 and 7000 years ago. Some regional domesticated crops included two pulses and two millet-grasses. Because the evidence indicates the earliest domesticated crops were native species, it is possible that independent domestication did occur in the southern region of India (Fuller et al., 2004).

In India, plant cultivation and domestication may have begun separately in five different regions: South India, Orissa, the Middle Ganges, Saurashtra, and the Himalayan foothills of the Punjab Region. However, there is no unambiguous evidence for local domestication or any evidence that domestication occurred without introduced crops (Fuller, 2011). We have seen that northeast India reaches into the primary region of rice domestication and that mango and orange are likely to have been brought into the domus there. Sesame, pigeonpea, eggplant, guar, several minor millets, several pulses, and some tubers were domesticated locally.

The processes of domestication swept across Asia from the Levant and Anatolia to the Pacific Ocean. I can no longer find evidence for centers of origin except for individual crops, and these are often multiple.

Conclusions

The Far East represents a highly diverse set of environments, and crop domestication follows a noncentric pattern similar to Africa. Agriculture was apparently established somewhat later than the Middle East. However, a mosaic of Mesolithic cultures and unique tools developed in regional and local settings. Many plants were domesticated from local flora, from India to Japan and throughout the Pacific Islands. Northern China advanced a millet agricultural system while those in the south were dominated by rice culture. Crops developed in the Far East, such as rice, soybean, banana, sugarcane, many fruits, spices, and tea, are very important to present-day agriculture.

References

Allen, J., Gosden, C., & White, J. P. (1989). Human Pleistocene adaptations in the tropical island Pacific: Recent evidence from New Ireland, a Greater Australian outlier. *Antiquity, 63*, 548–561. https://doi.org/10.1017/S0003598X00076547

Balbaligo, Y.E. (2015). *Ceramics and social practices at Ille Cave, Philippines* (Doctoral dissertation). University College London, London.

Barker, G., & Richards, M. B. (2013). Foraging–farming transitions in island Southeast Asia. *Journal of Archaeological Method and Theory, 20*, 256–280. https://doi.org/10.1007/s10816-012-9150-7

Barrau, J. (1965). L'humide et le sec, an essay on ethnobotanical adaptation to contrastive environments in the Indo-Pacific area. *The Journal of the Polynesian Society, 74*, 329–346.

Bellimod, P. (2004). The origins and dispersals of agricultural communities in Southeast Asia. In I. Glover & P. Bellwood (Eds.), *Southeast Asia: From prehistory to history* (pp. 21–40). London: Routledge Curzon.

Bellwood, P. (1985). *Prehistory of the Indo-Malaysian archipelago*. Orlando, FL: Academic Press.

Chang, K.-C. (1986). *The archaeology of ancient China* (Vol. 4). New Haven, CT: Yale University Press.

Chang, T. T. (1989). Domestication and the spread of the cultivated rices. In D. R. Harris & G. C. Hillman (Eds.), *Foraging and farming: The evolution of plant exploitation* (pp. 408–417). London: Unwin Hyman.

Christie, A. (1968). *Chinese mythology*. London: Hamlyn.

Cook, O. F. (1939). *A new palm from Cocos Island collected on the presidential cruise of 1938. Smithsonian Miscellaneous Publication No. 98*. Washington, DC: Smithsonian Institute.

Coursey, D. G. (1976). Yams. In N. W. Simmonds (Ed.), *Evolution of crop plants* (pp. 70–74). London: Longman.

de Oviedo y Valdes, G. F. (1944). *Historia General y Natural de las Indias*. Asuncion, Paraguay: Guarania.

Denham, T., Haberle, S., & Lentfer, C. (2004). New evidence and revised interpretations of early agriculture in Highland New Guinea. *Antiquity, 78,* 839–857. https://doi.org/10.1017/S0003598X00113481

Fuller, D. (2011). Finding plant domestication in the Indian subcontinent. *Current Anthropology, 52,* S347–S362. https://doi.org/10.1086/658900

Fuller, D., Korisettar, R., Venkatasubbaiah, P., & Jones, M. K. (2004). Early plant domestications in southern India: Some preliminary archaeobotanical results. *Vegetation History and Archaeobotany, 13,* 115–129. https://doi.org/10.1007/s00334-004-0036-9

Glover, I. (1986). *Archaeology in eastern Timor, 1966–67*. Canberra, Australia: Department of Prehistory, Research School of Pacific Studies, Australian National University.

Golson, J. (1984). New Guinea agricultural history: A case study. In D. Deneen & C. Snowden (Eds.), *A time to plant and a time to uproot: A history of agriculture in Papua New Guinea* (pp. 55–64). Boroko, Papua New Guinea: Institute of Papua New Guinea Studies.

Groube, L. (1989). The taming of the rain forests: A model for late Pleistocene forest exploitation in New Guinea. In D. R. Harris & G. C. Hillman (Eds.), *Foraging and farming: The evolution of plant exploitation* (pp. 292–304). London: Unwin Hyman.

Gunn, B. F., Baudouin, L., & Olsen, K. (2011). Independent origins of cultivated coconut (*Cocos nucifera* L.) in the Old World tropics. *PLoS One, 6*(6), e21143. https://doi.org/10.1371/journal.pone.0021143

Harlan, J. R. (1975). Our vanishing genetic resources. *Science, 188*(4188), 618–621.

Higham, C., & Maloney, B. (1989). Coastal adaptation, sedentism, and domestication: A model for socio-economic intensification in prehistoric Southeast Asia. In D. R. Harris & G. C. Hillman (Eds.), *Foraging and farming: The evolution of plant exploitation* (pp. 650–666). London: Unwin Hyman.

Higham, C. F. W. (1976). Reply to K.L. Hutterer. *Current Anthropology, 17*(2), 221–242.

Ho, P.-T. (1969). The loess and the origin of Chinese agriculture. *The American Historical Review, 75*, 1–36. https://doi.org/10.2307/1841914

Ho, P.-T. (1974). *The cradle of the East: Enquiry into the indigenous origins of techniques and ideas of Neolithic and early historic China 5,000–1,000 BC.* Chicago, IL: University of Chicago Press.

Hort, A. (1916). *Theophrastus enquiry into plants.* Cambridge, MA: Harvard University Press.

Hutterer, K. L. (1976). An evolutionary approach to the Southeast Asian cultural sequence. *Current Anthropology, 17*(2), 221–242. https://doi.org/10.1086/201711

Hutterer, K. L. (1983). The natural and cultural history of southeast Asian agricultural ecological and evolutionary considerations. *Anthropos, 78*, 169–212.

Jones, R., & Meehan, B. (1989). Plant foods of Gidjingali: Ethnographic and archaeological perspectives from northern Australia on tuber and seed exploitation. In D. R. Harris & G. C. Hillman (Eds.), *Foraging and farming: The evolution of plant exploitation* (pp. 120–135). London: Unwin Hyman.

Kirch, P. (1982). Advances in Polynesian prehistory: Three decades in review. In F. Weindorf & A. E. Close (Eds.), *Advances in world archaeology* (Vol. 1, pp. 51–97). New York, NY: Academic Press.

Komarov, V. L. (1934). *Flora of the USSR* Akad. Nauk. Leningrad. (N. Landau, trans.). Jerusalem, Israel: Israel Program for Scientific Translations Ltd.

Kovach, M. J., Sweeney, M. T., & McCouch, S. R. (2007). New insights into the history of rice domestication. *Trends in Genetics, 23*, 578–587. https://doi.org/10.1016/j.tig.2007.08.012

Koyama, S., & Thomas, D. H. (1982). *Affluent foragers: Pacific coasts east and west.* Osaka, Japan: National Museum of Ethnology.

Laufer, B. (1919). *Sino-Iranica. Field Museum of Natural History Publication 201, Anthropology Series.* (Vol. 15). Chicago, IL: Field Museum.

Li, H. L. (1970). The origin of cultivated plants in Southeast Asia. *Economic Botany, 24*, 3–19. https://doi.org/10.1007/BF02860628

Molina, J., Sikora, M., Garud, N., Flowers, J. M., Rubinstein, S., Reynolds, A., . . . Purugganan, M. D. (2011). Molecular evidence for a single evolutionary origin of domesticated rice. *Proceedings of the National Academy of Sciences of the United States of America, 108*, 8351–8356. https://doi.org/10.1073/pnas.1104686108

Moore, D. R. (1976). Reply to Hutterer. *Current Anthropology, 17*(1), 221–242.

Oka, H. I. (1988). *Origin of cultivated rice.* New York, NY: Elsevier.

Pope, G. (1989). Bamboo and human evolution. *Natural History, 98*, 48–56.

Pyramarn, K. (1989). New evidence on plant exploitation and environment during the Hoabinhian (late stone age) from Ban Kao caves, Thailand. In D. R. Harris & G. C. Hillman (Eds.), *Foraging and farming: The evolution of plant exploitation* (pp. 282–291). London: Unwin Hyman.

Rawal, H. C., Amitha Mithra, S., Arora, K., Kumar, V., Goel, N., Mishra, D., . . . Solanke, A. U. (2018). Genome-wide analysis in wild and cultivated *Oryza* species reveals abundance of NBS genes in progenitors of cultivated rice. *Plant Molecular Biology Reporter, 36*(3), 373. https://doi.org/10.1007/s11105-018-1086-y

Renfrew, J. M. (1969). The archaeological evidence for the domestication of plants: Methods and problems. In P. J. Ucko & G. W. Dimbleley (Eds.), *The domestication and exploitation of plants and animals* (pp. 149–172). Chicago, IL: Aldine.

Simmonds, N. W. (1976a). Bananas. In N. W. Simmonds (Ed.), *Evolution of crop plants* (pp. 211–215). London: Longman.

Simmonds, N. W. (1976b). Sugarcanes. In N. W. Simmonds (Ed.), *Evolution of crop plants* (pp. 104–108). London: Longman.

Spriggs, M. (1989). The dating of the island southeast Asian Neolithic: An attempt at chronmetric hygiene and linguistic correlation. *Antiquity, 63*, 587–613. https://doi.org/10.1017/S0003598X00076560

Wafer, L. (1934). *A new voyage and description of the Isthmus of America.* Ser. 2. (Vol. *73*). Oxford, UK: Hakluyt Society.

White, J. P., & Connell, J. F. (1982). *A prehistory of Australia, New Guinea and Sahul.* Sydney, Australia: Academic Press.

Wu, G.A, Prochnick, S., Jenkins, J., Salse, J. Hellsten, U. Murat, F. Perrier, X. ... Rokhsar, D. (2014). Sequencing of diverse mandarin, pummelo and orange genomes reveals complex history of admixture during citrus domestication. *Nature Biotechnology, 32*, 656–662. https://doi.org/10.1038/nbt.2906

Yang, S. T. (1998). The Neolithic cultural relation between Guangdong and its surrounding regions. In S. T. Yang (Ed.), *Lingnan Wenwu Kaogu Lunji Lingnan relic archaeology papers* (pp. 271–281) [In Chinese]). Guangzhou, China: Guangdong Ditu.

Yi, S., Lee, J. J., Kim, S., Yoo, Y., & Kim, D. (2008). New data on the Hoabinhian: Investigations at Hang Cho Cave, Northern Vietnam. *Indo-Pacific Prehistory Association Bulletin, 28*, 73–79. https://pdfs.semanticscholar.org/f193/ad3a2bd8aba5cfe8cfa3da6fcec0ba865471.pdf

Zhu, Z., Dennell, R., Huang, W., Wu, Y., Qiu, S., Yang, S., . . . Ouyang, T. (2018). Hominin occupation of the Chinese Loess Plateau since about 2.1 million years ago. *Nature, 559*, 608–612. https://doi.org/10.1038/s41586-018-0299-4

11

The Americas

Then the Maker and Creator asked them: "What do you think of your estate? Do you not see it? Are you not able to hear? Is your speech not good and your manner of walking about? Behold! Contemplate the world; see if the mountains and valley appear! Try, then, and see!

Mayan creation myth: *Popol Vuh*
Recinos (1947; J. R. Harlan translation)

Archaeology

Agricultural complexes also arose in the New World. An impressive array of Native American plants was domesticated by Native American people and agricultural systems eventually evolved sufficiently to

Source: Patricia J. Scullion

Harlan's Crops and Man: People, Plants and Their Domestication, Third Edition.
H. Thomas Stalker, Marilyn L. Warburton, and Jack R. Harlan.
© 2021 American Society of Agronomy, Inc. and Crop Science Society of America, Inc.
Published 2021 by John Wiley & Sons, Inc.
doi:10.2135/harlancrops

support the civilizations of Chavin, Olmec, Maya, Aztec, Inca, and others. Some extreme diffusionists have maintained that these developments were not independent of the Old World, and that the idea of cultivating plants was transmitted across the Pacific or Atlantic Oceans at a very early time (Riley et al., 1971). Basic to the argument is the implication that the Native Americans were incapable of innovation and had to be taught how to cultivate plants by people who had already invented or discovered the arts of agriculture. As we shall see, native peoples in the Americas were among the most skillful of all plant domesticators and it is difficult to understand why it should be thought that they were devoid of originality with respect to plant manipulation.

Asian and African plants are conspicuously missing from American crop complexes. Most damaging to the diffusionist argument, however, is the time required to develop an indigenous American agriculture. As we have already seen, the advantages of growing plants on purpose are not conspicuous at the beginning and the differences between intensive gathering and cultivation are minimal. It is difficult to imagine that a few sailors from Asia or Africa could easily induce people to take up practices that would not achieve a developed agriculture for several thousand years. It would appear much more likely that indigenous Americans began the process of domestication for about the same reasons as people of the ancient Near East, Africa, China, or Southeast Asia.

The archaeological record of man in the Americas is not nearly as long as that in the Old World. No remains of *Homo erectus*, Acheulean hand axes, or Chinese choppers have been found. The conventional wisdom has been that the Americas were peopled by big game hunters who crossed the Bering land bridge in the late Pleistocene when sea levels were low and there was a nearly ice-free corridor between the eastern and western glaciers down through Canada to ice-free land to the south. The theory has had problems in the testing, but new archaeological evidence is helping refine the picture.

There were, indeed, big game hunters to the south of the ice sheets. They manufactured a distinctive projectile point called "Clovis" after a site in New Mexico where a point was found imbedded in the skeleton of a kind of bison, now extinct. The date is about 12,000 years ago, and Clovis points have been recovered over an enormous range of both North and South America.

The people were specialists in big game mammals, hunting mammoth, mastodon, horse, camel, giant bison, ground sloth, and so on, and the culture seems to have died out as these animals became extinct. But, Clovis points have not been found in Siberia or Beringia and evidence of pre-Clovis people south of the ice is difficult to prove. Stanford (1983) has a short list of possible sites, but there are problems in dating or stratigraphy with each one of them. Meadowcroft, a cave on the border between Ohio and Pennsylvania, now has several dates in agreement at some 21,000–22,000 years ago, but the flora and fauna do not seem to fit the time range very well (Shutler, 1983). The Pleistocene animals, which later became extinct, and a boreal flora are missing.

When the earliest people arrived in the Americas has been greatly debated and may never be solved. The earliest record of man in the New World is Pebble flake industries, which are reported from Eastern Brazil and dated to approximately 30,000–35,000 years ago (Bryan, 1983; Schmitz, 1987). The Huaca Prieta site (600 km north of Lima, Peru) had evidence of early human occupation by 15,000 years ago (Dillehay et al., 2017); and Monte-Verde, southern Chile at 14,220 years ago, and possibly as early as 18,500 years ago (Dillehay et al., 2015a, 2015b). In North America, man was present at least by 24,000 BP at the Bluefish Caves in the Yukon Territory of Alaska (Bourgeon et al., 2017). By 12,000–14,000 BP, man had apparently spread across the entire North American continent and south into Central America. The early human skulls in both North and South America were of long-headed people, the later skulls short-headed (Carter, 1980). Asian Mongoloids are short-headed and have the highest known frequencies of *B* and *Rh2* genes, while both are low to nonexistent in Australians and Native Americans (Shutler, 1983). Still, *absolute proof* of pre-Clovis man in America is elusive (Dincauze, 1984).

In North America, the Bluefish caves remains of late Pleistocene fauna are preserved in excellent condition and include: mammoth, horse, bison, sheep, wapiti, and caribou, together with human artifacts, especially worked bone, horn, and ivory. Upper layers of these caves had an impoverished Holocene fauna and no evidence of humans. Archaeology has not supported the conventional wisdom of early man in the New World very well, not that people did not cross over Beringia, but there are possible or even likely alternatives to the scenario. Bednarik (1989) proposed a theory to account for both early Americans and the settling of Australia at 40,000 years ago or earlier. There could have been, he wrote, a Pleistocene coastal seafaring people who settled both Australia and South America, and rising sea levels have since covered the evidence. How and when they got to South America remains to be worked out. Lathrap (1977) has suggested an early colonization from Africa.

Our attention is directed toward the end of the Pleistocene and start of the Holocene for crop domestication. The punctuation period for North America is now well dated at 11,200–10,800 years ago. Large mammal extinction was more or less completed by 10,000 years ago with loss of some 33% genera or 70% of the large mammals, including camel, horse, mastodon, mammoth, ground sloth, a large cat, and others (Bonnichen et al., 1987). In South America, the Amazonian rainforest was not where it is now; it was restricted to several relatively small refuges around the perimeter of its present range (Haffer, 1969; Meggers et al., 1973). Rapid changes in fauna and flora followed and the scene was set for plant domestication. A few of the more important sites will be mentioned (Figure 11.1).

Guitarrero Cave in Peru is located in an Intermontane Valley on the west slope of the Andes at 2580 m elevation. It was excavated by Thomas Lynch (1980) and has a record over 12,000 years long. At the bottom level, a bifacial "knife" was found dated to approximately 12,500 years ago. Some projectile points were found above this and plant remains were found in complex II, dating 10,600–7600 BP. Plants include *Phaseolus vulgaris, P. lunatus* L. *Oxalis, Capsicum, Solanum asperolanatum* Ruiz & Pav., *Cucurbita* spp., *Inga,* and some grasses, perhaps for bedding. The common bean and pepper were found in Alton Ha, Belize, dated 10,600–10,000 BP and are domesticated types. The upper levels of II, 8600–7500 BP, had two types of beans, *Oxalis* spp. *and Solanum* spp. Complex IIa also had textiles, the second oldest in the New World. The oldest found so far is from Danger Cave, Utah, radiodated about 11,600 BP. It seems that beans were definitely domesticated before 10,000 BP. Wild beans and wild peppers are found on the eastern slopes of the Andes and were probably domesticated there and introduced to the west slope later.

Pachamachay Cave in the central puna of Peru at some 4000 m elevation has a record from before 11,000 BP. The earlier dates are rather hazy, but after 11,000 BP become rather clear. The site was excavated in 1974–1975 by John Rick (1980) and the abundant plant materials studied by Deborah Pearsall. The people were heavy exploiters of vicuña (a wild South American camelids). Domesticated beans and peppers appeared by 10,500 BP, confirming the finds of Guitarrero Cave. Sites on the coast yield abundant plant remains in superb condition, but at later dates. Skeletons were also well preserved and show that the earlier people were of the long-headed type, but by ceramic times, short-headed types had become dominant (Keatinge, 1988).

Guilá Naquitz Cave in Oaxaca, Mexico, was excavated by Kent Flannery and the site report published in 1986 (Flannery, 1986). The record covers the time range from hunting–gathering to farming. The foragers roasted maguey, ate cactus fruits, extracted syrup from mesquite pods, leached acorns, used wild onions and wild bean flowers; they had maguey-fiber sandals, the atlatl, fire drills, coiled baskets, net bags, snares, traps, and grinding stones.

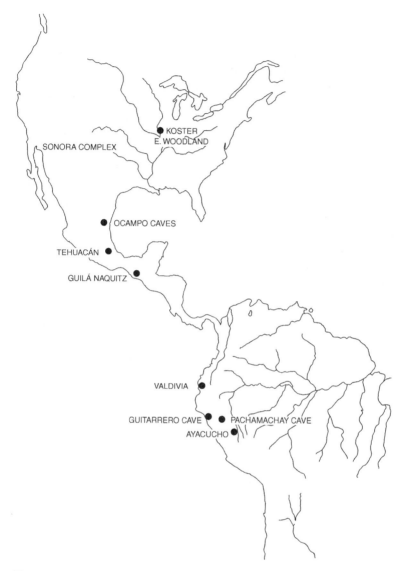

Figure 11.1 Examples of early sites of plant domestication in the Americas.

In the neighborhood of 10,000 BP, they too began to grow plants on a small scale, and again, beans and cucurbits were among the earliest domesticates. Some of the later sequences give clues about maize domestication. At least, teosinte pollen shows up around 9000 BP. The process seemed to be very slow and deliberate, and the small-scale efforts at gardening may not have been designed so much to increase the food supply as to make it more dependable.

There are other sequences in the Americas that help round out the picture (Table 11.1). The Ocampo caves in Tamaulipas, Mexico yielded bottle gourd as early as 9000 BP and squash of some kind perhaps as early. Beans appeared by 6000 BP. Remains of early cucurbits also show up at Koster, Illinois approximately 7000 BP, and an Eastern Woodland complex had developed by 5500–5000 BP. The complex began with gardens of squash and bottle gourd, to which were added sump weed (*Iva annua* L.) by 4000 BP, a local lambsquarters (*Chenopodium berlandieri* Moq.) before 3000 BP and sunflower by 3000 BP (Watson, 1989). Over the next few centuries, a canary grass (*Phalaris caroliniana* Walter), little barley (*Hordeum pusillum*),

Table 11.1 Some early American archaeological sites and associated cultivated plants found at the respective locations.

Date in millennia BC	Site	Plant remains
8.7-7.8	Guilá Naquitz, Oaxaca, Mexico	*Cucurbita pepo* (seed fragment)
8.5-8.0	Guitarrero Cave, Peru	*Phaseolus, Capsicum*
8.5	Pachamachay Cave, Peru	*Phaseolus, Capsicum*
8.2	Guitarrero Cave, Peru	*Phaseolus vulgaris, Oxalis, Capsicum, Solanum*
7.4-7.2	Guilá Naquitz, Oaxaca, Mexico	*C. pepo* (seeds and peduncles), *Lagenaria*
7.0	Ocampo Caves, Mexico	*Lagenaria*
5.5-5.5	Tehuacán, Puebla, Mexico	*Capsicum*
6.0-4.0	Coastal Peru	*Lagenaria*
5.7	Guitarrero Cave, Peru	*Phaseolus vulgaris, P. lunatus*
5.5	Tehuacán, Puebla, Mexico	*Lagenaria, Zea mays*
5.5	Poro Ayacucho, Peru	*Lagenaria*
5.0	Koster, IL, USA	*C. pepo*
4.0	Ocampo Caves and Puebla	*P. vulgaris*
3.5	Tehuacan, Puebla	*Gossypium, Amaranthus*
3.5-3.0	Eastern woodland complex, USA: MO, IL, IN, OH, MI, TN, KY, MS	*C. pepo, Lagenaria*
3.3	Real Alto (Valdivia phase), Ecuador	*Canna, Canavalia, Gossypium, Zea*
2.1	Eastern woodland, USA	*Iva, Chenopodium*

and a ragweed (*Ambrosia trifida* L.) were added to the complex. The early squash may have been domesticated locally from a wild race of *Cucurbita pepo* L. *subsp. ovifera* (L.) D. S. Decker var. *texana* (Scheele) Filov (Heiser, 1989). The eastern woodland complex was well developed before the arrival of Mexican crops such as maize, beans, tobacco and other kinds of squash.

Still another independent development took place in Sonora, Mexico, and the southwest United States. It involved maize, gourds, squash, beans, amaranth, and cotton from neighboring regions, but devil's claw and tepary bean were domesticated locally (Ford, 1981, 1985).

In the tropics, lowland infiltration of human activity into the rainforest was probably late and slow to develop because the undisturbed forest is an extremely difficult environment to exploit (Bailey et al., 1989; Piperno, 1989). Man-modified forests are much richer in exploitable plants, and the plants are much more concentrated than in climax situations. Piperno (1989) records evidence that in Panama, the hunter-gatherers were very few and had little impact on the region until about 7000 BP when maize-based agriculture took over. Developments in Amazonia were later and probably slower because of the size of the area. There are a number of forest plants that provide significant amounts of food: Brazil nut, cashew, peach palm, carrizo palm, among others, but the savanna plants are generally more important, and these must be adapted to forest conditions.

The general picture that emerges is one of widespread and early plant manipulation. Beans, cucurbits, and *Capsicum* peppers are generally among the first in each area regardless of the time range. These were grown on a small scale for a long time, with additional crops being added from time to time, one by one, and with seeming reluctance. Eventually, agricultural complexes evolved and fully agricultural economies became established. The complexes ranged from the midwestern United States to at least northern Argentina and evolved over several thousand years.

The Crops

Cereals

The major cereal of the Americas is maize, and the origins have recently become clear with the advent of DNA sequencing technologies. Putting to rest the many (and often contentious) discussions and theories of maize evolution, we now know that all maize is derived from one ancestral teosinte, *Zea mays* L. subsp. *parviglumis* H.H. Iltis & Doebley, specifically, from the central region of the Central Balsas River Valley in the Mexican state of

Guerrero (Doebley, 2004; Matsuoka et al., 2002). Additionally, despite theories put forward for independent domestication events in other places, driven mainly by the remarkable diversity of maize throughout the Americas, and especially the maize complexes of Meso-America, there was only one domestication event according to the DNA evidence. The vast phenotypic diversity is due to factors mentioned earlier (large effective population sizes, a plastic genome and, importantly, measurable gene flow from other teosintes) (Ross-Ibarra et al., 2009; Warburton et al., 2011). Further gene flow from tripsacums into maize has not been measurable, but has been seen between species of *Tripsicum* and between *Tripsacum* and teosinte species, which probably acted as a bridge to maize. We do know that the distribution of teosinte was much wider at one time than the present (Figure 11.2). It was found archaeologically in Tamaulipas, Mexico, where it does not occur today (Mangelsdorf et al., 1967). Specimens have been collected from sites where teosinte no longer occurs and are on file in herbaria (Wilkes, 1967). A tetraploid race has disappeared since 1910 from Ciudad Guzman, Mexico. All of these could have contributed to local maize landraces via gene flow.

Was some kind of teosinte once in South America? We do not know, but a diary entry of November 7, 1777 that was cited by José Celastino Mutis

Figure 11.2 Documented sites of teosinte distribution in Central America. *Source:* Adapted from Wilkes (1967).

(1957) mentions an inflorescence brought to him by a friend who had just been for a walk. They both recognized it as some kind of maize, but different from the cultivated kind. Mutis called it *Maicillo Cimarron* (*Zea sylvestris* Lam.). He was living at the time at Las Minas del Sapo, near Ibagué, Colombia. A month later, he wrote a brief Latin description, which is on file at the Jardin Botánico in Madrid. The handwriting, spelling, and Latin are all difficult. There is no specimen. The fact that the terminal inflorescence had only male flowers would seem to indicate the plant was, indeed, a *Zea* and not *Tripsacum*. On two visits to Ibagué, I was unable to find any such plant. I visited the Jardín Botaníco to photograph the original description in Mutis' handwriting. But firm conclusions are difficult to reach. Mutis was the most prominent botanist in South America at the time, and one is inclined to believe he could tell wild maize from domesticated maize.

All populations of teosinte, which I have seen, have both long branch and short branch types of plants. Each small raceme has several fruit cases containing one female fertile spikelet and a reduced sterile one. Each fertile spikelet has one fertile floret and one reduced sterile one. The male spikelets in the tassel are also in pairs, both are fertile, and both florets are pollen-bearing as well.

While the ear of maize looks very different from the small, fragile racemes of teosinte, all the flower parts and structures are present in both. Simple mutations in only three traits, one known to be encoded by a single gene, are needed to change teosinte racemes into something very similar to a maize ear: disarticulation of the teosinte ear for dispersal; two ranks of seeds in teosinte but four in maize; and removal of the hard fruit case around each teosinte kernel. This last trait is controlled by *teosinte glume architecture* (*tga1*), a mutation that removes the shell in maize. In addition, the *teosinte branched1* (*tb1*) gene by itself changed teosinte into more maize-like plant forms, as it changes the long lateral branches of teosinte, each tipped by male tassels, to the short lateral branches of maize, a few of which are tipped by ears. In addition, a limited number of other traits, many of which have been mapped to regions of the maize genome (reviewed in Tian et al., 2009), can produce all the changes required to convert teosinte into a primitive maize such as that found at Tehuacán. The long mystery regarding maize origins, which led to many contentious scientific exchanges, has been solved.

What would attract humans to a plant like teosinte, with its hard fruit cases and relatively modest yield? George Beadle ground up fruit cases, seed and all, made tortillas out of the flour and ate them. Several of us harvested the fruit cases to see what kind of yields we could get. My personal opinion is that teosinte was first used as a vegetable. Working in a field of teosinte on a hot day, I have many times broken off young, succulent ear branches and stuffed them in my mouth. They are tender, juicy, sweet, and refreshing. Very immature maize cobs have the same quality, and baby corn has been bred to yield multiple tiny ears popular in many Asian dishes. We have even found quids of chewed plant parts in archaeological sites. I believe teosinte was taken into the home garden and there the critical mutation occurred, possibly more than once. The full story of maize evolution is being documented archaeologically (Blake, 2006; Piperno & Pearsall, 1998; Reyna & Alvarez, 2009) and with the full suite of new DNA sequence-based tools, information on maize farming is now well attested in many Latin American countries by about 7000 BP (Grobman et al., 2012; Liu et al., 2015; Piperno, 1989; van Heerwaarden et al., 2011).

Other cereals domesticated in the Americas were very minor indeed. *Panicum hirticaule* was grown on a small scale in Sonora (Nabhan & de Wet, 1984). *Setaria parviflora* was apparently once domesticated in Mexico, but was abandoned when maize became available (Callen, 1967). In the southern Andes, *Bromus mango* É. Desv. was once grown; and was described rather hazily by early cronistas, but abandoned with the importation of European cereals. Additional grasses were grown or managed by natives of the southern California–Mexico border zone (Shipek, 1989). In the Eastern Agriculture Complex of the United States, Native Americans grew domesticated forms of wild rice (*Zizania palustris* L.), buckwheat (*Polygonum erectum* L.), little barley (*H. pusillum* Nutt.), and canarygrass (*P. caroliniana* Walter) (Pickersgill, 2007). However, these also were all virtually abandoned when maize arrived from Mexico via the U.S. Southwest, because none of these cereals could yield nearly as much as maize. Thus, except for maize, which was a real triumph of plant breeding (or the luckiest of accidents), the Native Americans did not maintain the grasses they had once planted and harvested on a considerable scale.

Beans

The annual grain legumes fared better than cereals. The five different bean species, including common bean, lima bean, runner bean, tepary bean, and year bean were domesticated, the first two at least twice independently. The small-seeded *P. lunatus* L., or seiva bean, was taken into the domus in Mexico, and the large-seeded lima was domesticated in South America. The ancestor of the common bean ranges from Mexico down the eastern slope of the Andes to Argentina and was domesticated twice, around 8000 years ago (Gepts, 1998). Lima bean, runner bean, and tepary bean were domesticated between 2000 and 3000 years ago. Genetic sequencing studies show two major complexes with extensive gene flow between them and many wild sister species (Rendón-Anaya et al., 2017).

Tomato

The domestication of the tomato presents a curious story. Wild tomatoes are found along the coasts of Ecuador and Peru. It was domesticated in Mexico, thousands of miles north of the range of the wild tomato. There is little evidence that the tomato was known to native people of South America in pre-Columbian times. There is no name for it in any of the South American languages, no tradition, no archaeological remains, nothing. The natives of Mexican had a name, *jitomatle* in Nahuatl (Aztec); they grew it, sold it in the markets, and prepared it in a variety of dishes. Jitomatle was well integrated into the culture of a number of tribes by the time of European contact; however, truly wild tomatoes are unknown in Mexico, and neither is the exact date of domestication.

Weedy tomatoes are common, however, in southern Mexico. They are common in and about the maize fields that are a part of the slash and burn bush–fallow rotation used in the region. They are also found along road-sides, in vacant lots, and growing on Maya ruins. It is thought that the weed race was the raw material for tomato domestication in Mexico (Jenkins, 1948), and that the weed race reached Mexico after agriculture had been well established. One clue to its late arrival is found in the Nahuatl names. Our word "tomato" is derived from the Native American word tomatl, but this originally applied to a cultivated *Physalis*. The Aztecs used the word *jitomatle* for tomato, implying that the *Physalis* had been domesticated first and the tomato came later as another kind of *tomatte*, in this case "sand tomato."

The wild and weedy races all have fruits with two locules, and some of the most popular cultivated kinds grown in tropical America are also two-loculed. They are thick-fleshed, firm, and have good flavor. Apparently, the tomato taken to Europe early on was a highly fasciated multiloculed type with prominent ridges. These are illustrated in 16th century herbals published in Europe. People spent 400 years trying to breed the ridges off when there were smooth types in Mexico all the time. An early introduction must also have been yellow, for such words as *pomodoro, pomme d'or, golden apfeln*, and so on, sprang up over Europe and are still in use in Italy and the former Soviet Union. Since the tomato obviously belongs to the nightshade family, it was also treated with suspicion and accused by some of being poisonous. Acceptance in Europe was generally slow. It was not until the 20th century that it became a major food crop on the world scene.

Squash

Squash and pumpkins do not quite make the select 30, but are very popular vegetables around the world. All belong to the genus *Cucurbita*, a New World taxon. Five species were domesticated: *C. maxima* Duchesne in South America and three in southern Mexico; *C. ficifolia* Bouché (a highland perennial species that reached the Andes in pre-Columbian times); *C. moschata* Duchesne; and *C. argyrosperma* C. Huber subsp. *argyrosperma*. The fifth species, *C. pepo* L., is the most widely grown and the most popular. Domestication of squash occurred in both North and South America, but it was first domesticated in Mexico approximately 8700 years ago (Ranere et al., 2009) and about 7000 years ago in South America (Pickersgill, 2007). The species *C. pepo* includes a remarkable array of fruit forms from Halloween pumpkins to crookneck, pattypan, and zuccini squashes. This one species seems to have been domesticated at least twice, once in northeastern Mexico and once in the Midwestern USA (Heiser, 1989). The progenitor bears the epithet *C. texana*, but is the same species as *C. pepo*. Maintenance of two epithets fits the usual bad taxonomy of cultivated plants.

The wild form and early domesticated types were pepo gourds. The fruits were relatively small with very little and very bitter flesh. It might have been grown for the seeds, which are edible and oil-rich, and for rattles and containers. Squash flowers are eaten in Mexico and are very popular. Some cultivars have been selected primarily for their blossoms. Adult corn rootworms are attracted by the Bitter Principle (a group of plant constituents linked only by their pronounced bitter taste) and gourds can be used as bait to help control the pest. The transformation of such an unpromising gourd into the tender, sweet crookneck, pattypan, zuccini, and pumpkins of today is, again, a transformation like that of the wild peanut to the modern form and teosinte to maize.

Sunflower

The sunflower (*Helianthus annuus* L.) appeared in the Eastern Woodland Complex in the Mississippi–Ohio watershed around 4000–5000 years ago (Blackman et al., 2011). There is still conflicting opinions on the evolution of the domesticated sunflower, which will probably be resolved with more genetic sleuthing, but its original source was probably farther west in the plains rather than in the woodlands, and there are weedy races that take advantage of human disturbance. All species of the complex are weedy and support a "dump-heap" origin. The wild and weed sunflowers have many branches and many small heads. The evolved domesticated races have a single stalk with a huge terminal head, and the achenes are several times the size of the progenitor races. The Native Americans had succeeded in producing this spectacular transformation before the arrival of Europeans. As a crop, it was not as important as maize, squash, or beans, and was grown in gardens rather than fields. It did not become a major crop on the world scene until after its introduction to Europe. There, freed from many of the diseases and pests of its native land, and subjected to intensive breeding and selection, it developed into one of the major food (oil) crops of the world. Sunflower has not fared as well in its homeland, but is of significant value in some U.S. states. It also suffers severe competition with other oilseed crops.

Peppers

The *Capsicum* peppers have been gratefully received by people around the world. The hot types are especially appreciated in the tropics, the sweet bell peppers are popular in temperate zones. Chili peppers derive their name from the Nahuatl language of the Aztecs, *"chīlli,"* with whom the plant originated. There are three entities involved (Pickersgill, 1989), *C. pubescens* Ruiz & Pav., *C. baccatum* L., and the *C. annuum* L.-*C. chinense* Jacq.-*C. frutescens* L. complex. The first two were domesticated in Bolivia and adjacent areas. The complex was domesticated at least twice, one type (*C. annuum*) in Mexico about 6500 years ago and the *C. chinense* type in Amazonia between 2500 and 4000 years ago (Kraft et al., 2014; Perry et al., 2007). The species and members within the complex are isolated by chromosome rearrangements.

The peppers that have captivated the world belong to the *C. annuum-C. chinense-C. frutescens* complex, and the very popular tabasco sauce uses a *C. frutescens* source from South America. This fermented product has been fondly accepted around the world. *C. annuum* types include bell peppers, sweet and Italian peppers, serrano, cayenne, paprika, and jalapeños. The *Capsicum* peppers are good sources of vitamins and impart anti-microbial properties to the foods in which they are cooked. The recent fad of breeding super-hot chiles have culminated in releases such as the bhut jolokia or ghost pepper (*C. chinense* × *C. frutescens*), and the California Reaper (*C. chinense*).

Peanut

The peanut was another triumph of plant breeding. The wild form, *Arachis monticola* Krapov. & Rigoini, occurs in Jujuy of northwestern Argentina and adjacent Bolivia on the eastern slopes of the Andes at moderate elevations.

This species originated from two diploid species [*A. duranensis* Krapov. & W. C. Greg. (A genome) and *A. ipaënsis* Krapov. & W. C. Greg. (B genome)] that spontaneously doubled in chromosome number, and polyploidy apparently increased the seed size. The wild forms usually have two small seeds in separate pods that are separated by a very long threadlike structure termed a peg. Morphologically, the wild-type is about as far from a jumbo or circus peanut as teosinte is from maize.

Although archaeological remains of the cultivated peanut (*A. hypogaea* L., $2n = 4x = 40$, AABB genomes) have been found in the Huarmey Valley in Peru dating to 5000 BP (Bonavia, 1982), the pod sample strongly resembles wild species. Radiocarbon dating also indicates that peanut was found in northern Peru in the Nancho Valley on the western slopes of the Andes that dated 7840 BP (Dillehay et al., 2007). However, these regions are far removed from the natural regions where wild peanuts grow and the most likely area of domestication was the Chaco lowlands in southern Bolivia (Krapovickas & Gregory, 1994). The species differentiated into two subspecies (*hypogaea* and *Valencia*) which each have three botanical varieties (Stalker & Simpson, 1995). A primary center of diversity exists in Bolivia and there are five secondary centers in five other geographic regions of South America (Wynne & Coffelt, 1982).

Root and Tuber Crops

Several wild roots or tubers were harvested by Native Americans in North America, but none were domesticated there. In tropical America, roots and tubers were main staples, equal or surpassing the importance of cereals in many places, and three root and tuber complexes evolved (Hawkes, 1989). In the lowlands, the main ones were cassava, sweet potato, a yam (*Dioscorea trifida* L.f.), and an aroid (*Xanthosoma*; also known as malanga or the new cocoyam) (Table 3.2). Cultivated cassava originated from its wild progenitor, *Manihot esculenta* subsp. *flabellifolia* (Pohl) Cif., by 7000 BP in the Amazon basin (Olsen & Schaal, 1999). Cassava, along with sweet potato, have become major crops on the world scene and supply food for many millions of people in Asia and Africa as well as the Americas. Yams of various species are popular around the world, but in some places they have been replaced by the faster growing sweet potato. The aroid malanga is popular in the Caribbean, Central America, and certain parts of Africa and Asia and is well suited to the wet tropics and the forest zones.

In the high Andes, a complex evolved adapted to the cold, rigorous conditions of the puna and adjacent valleys. The group includes the potato (*Solanum tuberosum* L.), a major world crop, while oca, ulluco, añu, and maca are still locally important but have not received much acceptance elsewhere. Cultivated potato was domesticated 8000–10,000 BP from one or more members of the *S. brevicaule* complex ($2n = 2x = 24$) native to the Andes of southern Peru (Spooner et al., 2005).

An intermediate group adapted to warm temperate climates includes: *Arracacia*, which is also grown in Brazil and the Caribbean; *Canna*, which is only grown as a minor food crop known as achira; *Polymnia* (synonymous to *Smallanthus sonchifolius* or locally as yacón), which has become popular in Japan for its antihyperglycemic properties among diabetic and health-conscious consumers; *Mirabilis expansa* (mauka or chago), a relative of the popular ornamental 4-o'clocks that was an important stable to the Incas; and some three species of *Pachyrrhizus*. The *Pachyrrhizus* or jicama of Mexico is extremely popular as a snack food in Southeast Asia and Oceania as well as tropical America. Altogether, some 25 species of root and tuber crops were domesticated, forming the largest group of this class in the world, and three of them appear on the select list of 30 major food plants (Table 2.1). All are derived from savanna zones with long dry seasons.

Sweet Potato

The origin of the sweet potato [*Ipomoea batatas* (L.) Lam] was unknown until molecular genetic evidence began to clarify the story (Roullier et al., 2013a, 2013b). It has a chromosome number of $2n = 90$ and thought to be an autohexaploid. The genus *Ipomoea* is a large one found in both the Old and New Worlds. Nishiyama (1971) implicates *I. trifida* L. from Mexico as the most likely progenitor, but the bulk of the evidence suggests domestication occurred in South America and most likely northern South America. The oldest reported archaeological finds are from Tres Ventanas in Chilca Canyon, Peru, and said to date from approximately 10,000 BP (Engel, 1984). The site is too high for the crop to have been grown locally, and the tubers must have been brought in from lower elevations from either side of the Andes. Other finds are considerably later, mostly from the second millennium BC and one at Ancon-Chillon on the Peruvian coast in the third millennium BC (Hawkes, 1989). Molecular evidence concurs that domesticated sweet potato originated from the ancestor it shares with *I. trifida,* but possibly from at least two distinct autopolyploidization events followed by gene flow between the two events and from wild *I. batatas* populations (Roullier et al., 2013a, 2013b).

In the 1770s, the Cook expeditions found the sweet potato on Easter Island, Tahiti, New Zealand, and Hawaii. The find in Tahiti seemed to be a single garden, suggesting a recent introduction, but it was well established and an important crop in New Zealand and Hawaii. The possibilities are: (a) the Polynesians sailed to American shores and returned with it, (b) Native Americans sailed to Polynesia where it was accepted and distributed by Polynesians, and (c) the crop was introduced into Polynesia one or more times by Europeans during the two and half centuries between Magellan and Captain Cook (Barrau, 1957).

Perhaps more interesting is the discovery during and after World War II that people in the highlands of New Guinea had developed a genuine sweet potato culture. It was and is the staple of the people and is also fed to pigs, which are the primary source of protein. Archaeological evidence indicates that a population explosion occurred in the highlands after its introduction. Increased populations may have been enhanced by the fact that the highlands are above the malarial belt where this population leveler does not operate. This has been called the "Ipomoea Revolution."

Dating the arrival of the sweet potato to the Pacific has not been easy. Several archaeological finds have turned up but always tantalizingly close to the time of Magellan. In much of the Pacific, the sweet potato was introduced together with the word *kumara* or variations of it. In the Philippines, where the crop is important, a Caribbean word, *camote*, came with it and the crop was clearly introduced by the Spanish. Kumara is a South American word, but an obscure one (Yen, 1974).

Cotton

Two diploid species of cotton, *Gossypium arboreum* L. and *G. herbaceum* L., were domesticated in the Old World, but now constitute less than 1% of world production, which is dominated by the American tetraploid species, upland cotton, *G. hirsutum* L., and the long-staple type *G. barbadense* L. The first was domesticated in southern Mexico 5000 BP from a wild *G. hirsutum* but a severe domestication bottleneck left the cultivated species with limited genetic diversity (Fang et al., 2017; Wendel et al., 1992). *Gossypium barbadense* was domesticated on the west coast of South America in Ecuador and Peru (Hu et al., 2019) by at least 4400 BP, but was introduced early on into Brazil, where selection helped shaped the species. Gene flow between the two tetraploid species has been shown (Fang et al., 2017). The bulk of the cotton grown worldwide is of the Mexican species, and the leading producers are China, India, the United States, Pakistan, and Brazil. The United States, Egypt, Sudan, China, India, Australia, Israel, Peru, Tajikistan, Uzbekistan, and Turkmenistan produce most of the *G. bardadense* long-staple cotton.

The New World cottons have AADD genomes; the A is homologous to the genomes of Old World *G. herbaceum* L. and *G. arboreum*. The archaeological remains of cotton are found on the Gulf Coast of southern Mexico, approximately 5500 BP, and on the central coast of Peru, 4500 BP (Phillips, 1976). Wild forms and early domesticates were perennial. Annual types evolved in recent historical times as cotton culture was taken into temperate zones. Cotton was grown mostly for home use until Whitney invented the saw gin in 1793. Mechanization of spinning and weaving followed, and the cotton industry became a major component of the industrial revolution; today, it is also a major oilseed crop. Cotton is an ancient crop, but its role as a major crop on the world scene is relatively new.

The presence of an A genome in American tetraploid cottons has raised speculation about early human transport across the oceans. Cytologically, the American A genome is closer to the African *G. arboreum* L. than to the Asian *G. herbaceum*, differing from the first by one chromosome interchange and from the second by two exchanges. No linted diploid has ever been found in the Americas. The American diploids are found only west of the Andes. There are five species of tetraploids, including the two domesticated ones already discussed, a relict *G. mustelinum* Miers ex G. Watt of northeast Brazil, *G. tomentosum* Nutt. ex Seem. of Hawaii, and *G. darwinii* G. Watt found only on the Galapagos islands. All have cytologically identical A genomes. Both of the domesticated American cottons have weedy races that do well without cultivation. The fact that there has been speciation to such a degree after the initial allotetraploid was produced argues strongly for an ancient origin of the tetraploid cottons, which has been placed by sequencing data to around 1.5 million BP (Zhang et al., 2015).

Tobacco

Use of tobacco in prehistoric times was confined to the New World and Australia, as far as we know. The narcotic was widely used in the Americas, with perhaps some 10 species involved. The most important, however, were *Nicotiana tabaccum* L. in South and Central America; *N. rustica* L. east of the Mississippi, northern Mexico, and the Caribbean; and *N. quadrivalvis* Pursh (syn. *N. bigelovii* S. Wats), *N. attenuate* Torr. ex S. Watson, and *N. trigonophylla* Dunal in western North America. The tobacco most important in commerce today is *N. tabaccum*, but *N. rustica* is grown on a relatively small scale primarily for insecticides based on nicotine, but also used as a smoking product. Neither of these two are known in the wild. Native American usage includes chewing (often with lime), snuff, and smoking pipes, or as cigars. Blowing smoke into a user's face was apparently common. Tobacco was important on ceremonial occasions.

Nicotiana tabacum is a tetraploid. Cytogenetic and molecular genetic evidence indicates the most probable diploid parents were *N. sylvestrus* Speg. as the maternal parent and *N. tomentosiformis* Goodsp. as the paternal parent, both endemic to small ranges on the eastern slope of the Andes, but not sympatric today. The cross occurred about 200,000 BP (Leitch et al., 2008); although the ranges are currently separated by about 3° of latitude, they may have overlapped at that time. The allotetraploid presumably appeared in or near the ranges of the progenitors, possibly in the Salta region of northwestern Argentina. No very early evidence of tobacco use has turned up archaeologically. Stone pipes are common enough, but mostly dating within the last 2000 years.

Nicotiana rustica is also a South American allotetraploid with no wild forms. The parents are probably *N. paniculata* and *N. undulata* of the central Andean region. Its extensive use in eastern North America and the Caribbean illustrates the ease of diffusion of such a plant. We know from recent history that a narcotic can spread very rapidly.

After tobacco was introduced to the Old World, it was taken up with alacrity in Europe, Asia, and Africa. Tobacco growing became a major industry in a short period of time, earning vast sums of money. The crop may not do much for the health of people, but has been important to the economy of many regions.

Rubber

Rubber is a new crop of the late industrial age. The native South Americans were aware of it and made some minor uses of rubber, but it was never important in either their agricultural or gathering economies until industrial demand around 1900 made exploitation of latex from wild trees profitable (Wycherley and Simmonds, 1976). The species of commerce is *Hevea brasiliensis* (Willd. ex A. Juss.) Müll. Arg., a tree of Amazonia; almost all of its range being south of the Amazon. Today, about 93% of commercial natural rubber is produced in Southeast Asia and the South Pacific, about 5% in Africa, and the rest in tropical America. Harvest from wild trees in Amazonia earns modest sums for forest-dwelling people, but Brazil imports rubber from Asia. Primary producers are (in descending order): Thailand, Indonesia, Malaysia, India, Vietnam, China, Sri Lanka, the Philippines, and Cambodia.

The primary constraint to plantation production in tropical America is South American leaf blight caused by *Microcyclus uleu* (Henn.) Arx. In the wild, trees are sufficiently scattered that epiphytotics do not build up, but under plantation conditions, the trees are devastated. Seeds from wild trees were collected from a single region near Boim on the Rio Tapajoz and westward by H. A. Wickham in 1876. He collected some 70,000 seeds from which 2800 seedlings were grown at the Royal Botanical Gardens in Kew, near London, and 2387 of them were sent to British-ruled Southeast Asia and to Indonesia. Most of the seedlings went to Sri Lanka, but some went to Malaysia, Singapore, and Bogor in Indonesia. In this manner, the species was introduced without the South American leaf blight.

Commercial production in Asia is based on the descendants of the Wickham collection. We thus have a clear case of modern domestication. The original stock was a sample of wild populations. Breeding and selection within this population has resulted in a several-fold increase in latex yield. Breeding in America has not had much success to date. One technique that might have potential is to graft a high-yielding clone to seedling stock and top graft this with disease-resistant clones. Each tree would consist of three genotypes and only the tapping panel would be of the high-yield constitution. The technique is feasible, but disease resistance keeps eroding populations due to changes in the races of the pathogen (Wycherley & Simmonds, 1976). The area of domestication of rubber and several other crops are shown in Figure 11.3.

Fruits, Nuts, and Ornamentals

The indigenous people of the Americas also delighted in tropical fruits and developed some of the most popular ones in the world. Consider such blue ribbon winners as pineapple, papaya, avocado, guava, passionfruit, cherimoya, and species of *Annona*. There are many others also of delightful flavor but not well known outside the regions of native production, for example, pepiño, lucuma, lulo, coconá, tomatillo, and so on. In temperate zones of the Andes and North America, wild cherry species including *Prunus serotine* Ehrh. and *P. viginiana* L., multiple plum species (*Prunus* spp.), bramble fruits (*Rubus*), strawberries, blueberries, cranberries, and American grape were exploited and some were domesticated. Cacao, the source of chocolate, was highly valued by natives of Meso-America, and the beans were used as money for tribute and exchange. Vanilla was also developed and cashew, Brazilnut, and pecan have become popular nuts around the world. The Native Americans provided a feast for the gods.

Mexican natives, especially, were fond of brilliantly colored flowers, and grew and sold them in the markets of their cities. Spanish cronistas were impressed by the attention devoted to commercial flower production, although

Figure 11.3 Examples of regions of domestication of several major world crops that were domesticated in the New World. Several crops were domesticated in both Mesoamerica and in South America, and two pepper species were domesticated in multiple regions of South America

ornamental and medicinal products were often confounded. Some of the most popular American flowers around the world include *Zinnia*, marigold (*Tagaetes*), *Fuchsia*, *Canna*, *Nicotiana*, and *Salvia*. The world is a more attractive and colorful place because of the efforts of Native American gardeners.

Forage Legumes

We are still finding useful wild plants in the American tropics and are domesticating them. One class being brought into the domus consists of forage legumes for pastures in the tropics. Improved tropical pastures are being developed in Australia, Africa, and Latin America. In general, the best grasses for the purpose come from Africa and the best legumes are American. Some of the most promising genera are from the family *Fabaceae,* which as nitrogen-fixing legumes, have the added advantage of being highly productive even in poor soils. Many of these include: *Centrosema,* also known as Centro and a productive and high-protein fodder gaining popularity in Australia and Southern Asia, as well as South American and the Carribean; *Stylosanthes,* several species or which are very important forage plants in tropical India and Africa, as well as South America; and *Desmodium,* or tick clover, several species of which produce secondary metabolites which repel some lepidopteran insects and the parasitic weed *Striga* when planted in maize or sorghum fields. Some *Desmodium* species also make good fodder and green mulch. Additional forage legumes come from the genus *Leucaena,* which include species with tree and shrub growing habits, several of which make good livestock fodder and one of which (*Lycopersicon esculentum*) is eaten in Mexico and known as guaje; *Arachis,* for example, *A. hypogaea, A. pintoi,* and *A. glabrata,* all from South America, are grown as nutritious forages in tropical environments worldwide; and *Macroptillium,* including *M. atropurpureum,* the purple bush bean, is a vigorously growing fodder but is considered a pest plant and banned in many countries.

Several collecting parties have been sent to various regions to gather plant species and forage breeding programs have been established to improve the best materials. As usual with legumes, there are problems with alkaloids and other poisonous substances, but progress is being made. New species being developed include *Cratylia argentea* (Desv.) Kuntze from the Amazon Basin and *Calliandra calothyrsus* Meisn. from the tropics of Central America. Who knows what other treasures might be found if we only looked?

Indigenous Americans as Biochemists

The high-elevation Andean complex of crops could not have been easy to develop; all, or nearly all of the species are poisonous. Either methods for detoxification had to be worked out for each plant or selection of strains with lower toxicity had to be made. Both were accomplished. Wild potatoes had toxic quantities of glycoalkaloids. Strains were selected that are relatively safe, but even today in areas far removed from the Andes, certain cultivars grown under certain conditions can be dangerous. In the Andes, strains selected for low toxicity frequently cross with wild or weedy strains and toxic tubers are produced. One must, somehow, live with the poisons.

Oca contains large quantities of oxalic acid, a very unpleasant substance when concentrated. It may be necessary to leach the tubers of some strains in running water for a matter of weeks followed by a freeze-drying operation to make them edible (Johns, 1989). Specially selected strains may be eaten with little or no processing. Maca and añu contain glucosinolates that, when damage to the tuber occurs, are converted enzymatically to isothiocyanates (mustard oil). The Native South Americans perceive these crops to affect human fertility, positively for women and negatively for men. There is probably some truth in the idea.

Ulluco, quinoa, and cañahua contain saponins and are bitter without processing. Mauka has flavenoids, saponins, and hemaglutinins and the cultivated lupin contains quinolizidine alkaloids as well as protease inhibitors. The proteinaceous poisons and inhibitors can be detoxified by cooking, but others require prolonged leaching, curing in the sun and/or freeze-drying, as well as cooking to make them edible.

Geophagy (clay eating) is common in the Andean region, and clay can bind with tannins, saponins, and alkaloids. It can also tie up iron and zinc and cause anemia. Clay eating is widespread throughout the world and is often practiced by people with no known problem with toxic foods. In the case of Native Americans of the high Andes, it might be a good precautionary measure. Selections for low toxicity have been made in all the edible tubers and in the forage plant lupin, but this often comes at some cost. The toxins have functions in protecting the plants from attacks of pests and pathogens. While leaching and freeze-drying is labor intensive, it works and so do the natural protectants. The Native Americans opted for a combination of detoxification techniques and selection for low toxicity. Much of the selection is based on taste (Johns, 1989), but simple selection for tuber size results in some reduction in toxicity.

In the lowlands, the most important tuber crop is cassava, or more properly manioc or mandioca. The word cassava was originally applied to the bread made from manioc. The word *yuca* is also commonly used. There are bitter and sweet types of manioc. All contain cyanogenic glycosides that can release deadly HCN, but the concentration in the sweet clones is much less than in the bitter ones. There are benefits in high glycoside content, however, and the Native South Americans have often selected for high toxicity. Bitter types are more resistant to attack by pathogens, pests, and animals that relish the sweet kinds. They yield more, and the cakes keep better. In regions where manioc is the primary food crop, bitter strains are preferred. Sweet types predominate where manioc is secondary to maize as a food crop.

There are two glycosides in the manioc tubers: linamarin and lotaustralin. They can be hydrolyzed by the enzyme linamarase to produce HCN, which is lethal in rather small quantities. Detoxification requires rupturing cells to bring the glycosides into contact with the extracellular enzyme. The tuber is peeled and grated and often put into a woven fiber press to express excess juice and make good contact. The meal must be incubated to allow enzymatic action to convert the glycoside to deadly but heat-labile HCN. An overnight incubation is usually enough. The poison is then driven off in cooking the cake. There are variations on the theme, but the essential steps are rupturing the cells, enzymatic action, and removal of the HCN. This is easily done by baking the cake, but HCN is volatile enough that drying in the sun can be adequate. Various fermentation procedures are also used. Sweet clones are low enough in toxicity that little or no processing is preformed; boiling or baking is considered sufficient. There are cases, however, of low-level chronic poisoning that ties up iodine and causes goiter as well as other health problems. These cases are usually in urban populations where the traditional ways have been forgotten.

Another important lowland tuber is *Xanthosoma*. It is usually grown in dooryard gardens, rather than extensive fields, and few figures on production are available. As a crop of the poor, its importance is often ignored. Wild and weedy races contain high levels of oxalic acid, often deposited in cells in bundles of needle-like crystals. A tiny bite of such a tuber can be extremely painful. Clones have been selected for low acid content, but prolonged leaching is required for others. There are many other cases where the people detoxified food (Johns & Kubo, 1988). One of the most famous is in the Pacific Northwest where the death camus, *Zygadenus*, was collected from mountain meadows and consumed on a substantial scale. Detoxification was by roasting in earth ovens.

The natives of the Americas were not only expert at detoxifying food, but apparently searched diligently through the flora for psychoactive drugs. According to Schultes and Hofmann (1979), the native Americans discovered and used about 130 plant species as compared to about 20 in the Old World. They list a sample of 66 belonging to 24 families. Of the 66, 11 are in Solanaceae, 9 in Leguminosae, 9 in Cactaceae, 8 in Agaricaceae (mushrooms), 4 in Compositae, and other families contributed one or two each. The list does not include coca, the most famous American drug of all. The psychoactive drugs were used in religious rite and ceremony, sometimes by the whole population, often by the shaman, medicine man, or by witches as the go-betweens. The drugs brought the user closer to his gods or to deceased ancestors. A drug from *Virola* (Myristicaceae) called epeña, for example, is taken as snuff and serves to transport the user to the spirit world. There is a sensation of flying and of actual communication with spirits. Two species in the Convolvulaceae yield LSD-like drugs. Mushrooms are a common and widespread theme in Native American art. *Triehocerus*, a psychedelic cactus, was carved on a stone slab at Chavín de Huantár, Peru, about 3300 BP (Schultes & Hofmann, 1979).

In addition to psychoactive drugs, extensive arrays of pharmaceuticals were developed for healing wounds and curing diseases. Early cronistas of Spain were much impressed by Native American pharmacology and quickly introduced medicinal plants to Europe and wrote herbals describing the plants and their uses. Aztec medicine was at least as good, and probably better, than European medicine at the time of contact, but the culture was destroyed so quickly that it was not studied in detail and much information was undoubtedly lost.

We are still losing opportunities. Surviving native American tribes, especially in the tropics, make use of many plant products unknown to modern science. They have cures for snake bite, insect bites of various sorts, centipede bites, scorpion stings, skin infections, internal parasites, head lice, colds, sores, inflammations of many kinds, and so on. It is said that the Tirió tribe of Suriname makes use of some 300 species of plants for drugs and medicine (Jackson, 1989). But, the forests are being cut down, the plants are disappearing, and the traditional lore is being forgotten. Our arrogance and prejudice may have cost us dearly in wasted opportunities to learn more about traditional medical botany and zoology.

People of the tropical forest were especially skilled and knowledgeable about fish and arrow poisons. Roots of several species of leguminous plants are used to stun fish. The roots are pounded at the edge of a forest pool and tossed in. Soon, fish come floating to the top, stunned but not necessarily killed. The natives gather as much as they wish, and the others recover. It is an efficient method of fishing. Fish poisons are not confined to the Leguminosae.

The most famous of the arrow or dart poisons is derived from the skin of certain tree frogs. There are over 100 species of the family Dendrobatidae; all are small, brilliantly colored frogs looking like little jewels. About half are poisonous, but only three species are regularly used. The most poisonous is the brilliant gold *Phyllobates terribilis* Myers, Daly, and Makin, with something like 20 times the toxicity of the other poisonous frogs used. The power of this poison is awesome. The area of usage is lower Central America and northern South America. Frogs made of gold are common in the art work of the region (Bainbridge, 1989).

The frog poisons are alkaloids and are the subject of some medical research. One group affects the neuromuscular system, another group acts as heart stimulants, and a third group acts as muscle relaxants and anesthetics. Plant-derived medicinals have become big business in the herbal remedy and supplement market, and some examples from the Americas include black cohosh (*Actaea racemosa* L.), bloodroot (*Sanguinaria canadensis* L.), devil's club (*Oplopanax* spp.), evening primrose (*Oenothera biennis* L.), and slippery elm (*Ulmus rubra* Muhl.), among others. Drugs of major economic and health importance have also been derived from plants of the Americas, including Tubocuranine, a well-known plant-derived drug used as an anesthetic in surgery derived from the curare vine (*Chondrodendron tomentosum* Ruiz & Pav.); quinine, a malaria and heart disease treatment from *Cinchona* species; scopolamine, for sedation; motion-sickness reduction derived from *Datura* and *Solananceae* species; and taxol, for the treatment of breast and other cancers, derived from *Taxus* species. How many other useful drugs have we failed to notice? Undoubtedly, we have a lot to learn from indigenous medicine, but time is running out to obtain the plants because native habitats are being displaced by cropland and urban development.

Conclusions

Native Americans were superb plant domesticators. Nine of their domesticates are on the elite list of 30 and account for about one-third of the edible dry matter of the entire list. Crops of the Americas have had profound effects worldwide. The sociopolitical impact of the potato on northern Europe, especially Ireland, Germany, and Russia was remarkable. Tomato, sweet pepper, and squash captured Mediterranean cuisine, while hot pepper transformed cuisine throughout the tropics and subtropics. Maize replaced sorghum in parts of Africa; cassava is more important in Africa than in South America where it originated. Peanut and cacao are basic to the economy of West Africa. American cotton is the premier natural fiber of the world. The sweet potato transformed highland New Guinea. In additional to the staples, the Native Americans contributed such delights as chocolate, vanilla, pineapple, papaya, and many other delicacies. They were extraordinarily adept at managing poisonous plants to make them edible and in searching the flora for psychoactive and healing plants. We may not owe great gratitude for tobacco and coca, but we must admit they have had profound effects on societies around the world.

Morphological change from wild to cultivated races were most spectacular in maize and peanut and not much less in sunflower. These changes are not exceeded by any other domesticates in the world. Geographically, the pattern resembles other parts of the world with activities of domestication moving on a broad front over vast regions. What once looked like a center of Mesoamerica has now been flanked by independent activities in northeastern and northwestern Mexico and southwestern and midwestern United States. The pattern is diffuse and follows an ecological map rather than one of "centers."

References

Bailey, R. C., Head, G., Jenike, M., owen, B., Rechtman, R., & Zechenter, E. (1989). Hunting and gathering in tropical rain forest: Is it possible? *American Anthropologist, 91*, 59–82. https://doi.org/10.1525/aa.1989.91.1.02a00040

Bainbridge, J. S. (1989). Frogs that sweat-not bullets but a poison for darts. *Smithsonian, 19*, 70–76.

Barrau, J. (1957). LL'enigme de la patate douce en Océanie. *Etudes d'Outre-Mer, 1*, 83–87.

Bednarik, R. G. (1989). On the Pleistocene settlement of South America. *Antiquity, 63*, 101–111. https://doi.org/10.1017/S0003598X0007561X

Blackman, B. K., Scascitelli, M., Kane, N. C., Luton, H., Rasmussen, D. A., Bye, R. A., . . . Rieseberg, L. H. (2011). Sunflower domestication alleles support single domestication center in eastern North America. *Proceedings of the National Academy of Sciences of the United States of America, 108*, 14360–14365. https://doi.org/10.1073/pnas.1104853108

Blake, M. (2006). Dating the initial spread of *Zea mays*. In J. E. Staller, R. H. Tykot, & B. F. Benz (Eds.), *Histories of maize: Multidisciplinary approaches to the prehistory, biogeography, domestication, and evolution of maize* (pp. 55–71). Amsterdam, The Netherlands: Academic Press.

Bonavia, D. (1982). *Precerámico Peruano. Low Gavilanes. Mar, desierto y oasis en la historia del hombre*. Lima, Peru: Corporación Financiera de Desarrollo S.A. Cofide and Instituto Arqueológico Alemán.

Bonnichen, R., Stanford, D., & Fastook, J. L. (1987). Environmental changes and developmental history of human adaptive patterns: The Paleoindian case. In W. F. Ruddiman & H. E. Wright (Eds.), *North America and adjacent oceans during the last deglaciation* (pp. 403–424). Boulder, CO: The Geological Society of America.

Bourgeon, L., Burke, A., & Higham, T. (2017). Earliest human presence in North America dated to the last glacial maximum: New radiocarbon dates from Bluefish Caves, Canada. *PLoS One, 12*, e0169486. https://doi.org/10.1371/journal.pone.0169486

Bryan, A. L. (1983). South America. In R. Shutler (Ed.), *Early man in the New World* (pp. 136–146). Beverly Hills, CA: Sage Publications.

Callen, E. O. (1967). The first New World cereal. *American Antiquity, 32*, 535–538. https://doi.org/10.2307/2694082

Carter, G. F. (1980). *Earlier than you think: A personal view of man in America*. College Station, TX: Texas A&M University Press.

Dillehay, T. D., Goodbred, S., Pino, M., Vásquez Sánchez, V., Rosales Tham, T., Adovasio, J., . . . Velchoff, N. (2017). Simple technologies and diverse food strategies of the Late Pleistocene and Early Holocene at Huaca Prieta, Coastal Peru. *Science Advances, 3,* e1602778. https://doi.org/10.1126/sciadv.1602778

Dillehay, T. D., Ocampo, C., Saavedra, J., Oliveira Sawakuchi, A., Vega, R., Pino, M., . . . Dix, G. (2015a). New archaeological evidence for an early human presence at Monte Verde, Chile. *PLoS One, 10,* e0141923. https://doi.org/10.1371/journal.pone.0141923

Dillehay, T. D., Ocampo, C., Saavedra, J., Oliveira Sawakuchi, A., Vega, R., Pino, M., . . . Dix, G. (2015b). Correction: New archaeological evidence for an early human presence at Monte Verde, Chile. *PLoS One, 10*(12), e0145471.

Dillehay, T. D., Rossen, J., Andres, T. C., & Williams, D. E. (2007). Preceramic adoption of peanut, squash, and cotton in northern Peru. *Science, 316,* 1890–1893. https://doi.org/10.1126/science.1141395

Dincauze, D. F. (1984). An archaeological evaluation of the case for pre-Clovis occupations. *Advances in World Archaeology, 3,* 275–323.

Doebley, J. (2004). The genetics of maize evolution. *Annual Review of Genetics, 38,* 37–59. https://doi.org/10.1146/annurev.genet.38.072902.092425

Engel, F. (1984). *Prehistoric Andean ecology: Man settlement and environment in the Andes.* New York, NY: Chilca Department of Anthropology, Hunter College, City University of New York Humanities Press.

Fang, L., Wang, Q., Hu, Y., Jia, Y., Chen, J., Liu, B., . . . Mei, G. (2017). Genomic analyses in cotton identify signatures of selection and loci associated with fiber quality and yield traits. *Nature Genetics, 49*(7), 1089–1098. https://doi.org/10.1038/ng.3887

Flannery, K. V. (1986). *Guilá Naquitz: Archaic foraging and early agriculture in Oaxaca, Mexico.* New York, NY: Academic Press.

Ford, R. I. (1981). Gardening and farming before A.D. 1000: Patterns of prehistoric cultivation north of Mexico. *Journal of Ethnobiology, 1,* 6–27.

Ford, R. I. (1985). *Prehistoric food production in North America. Anthropological Paper no. 75.* Ann Arbor, MI: Museum of Anthropology, University of Michigan.

Gepts, P. (1998). Origin and evolution of common bean: Past events and recent trends. *Horticultural Science (Prague), 33,* 1124–1130.

Grobman, A., Bonavia, D., Dillehay, T. D., Piperno, D. R., Iriarte, J., & Holst, I. (2012). Preceramic maize from Paredones and Huaca Prieta, Peru. *Proceedings of the National Academy of Sciences of the United States of America, 109,* 1755–1759. https://doi.org/10.1073/pnas.1120270109

Haffer, J. (1969). Speciation of Amazonian forest birds. *Science, 165,* 131–137. https://doi.org/10.1126/science.165.3889.131

Hawkes, J. G. (1989). The domestication of roots and tubers in the American tropics. In D. R. Harris & G. C. Hillman (Eds.), *Foraging and farming: The evolution of plant exploitation* (pp. 481–503). London: Unwin Hyman.

van Heerwaarden, J., Doebley, J., Briggs, W. H., Glaubitz, J. C., Goodman, M. M., Sanchez-Gonzalez, J. J., & Ross-Ibarra, J. (2011). Genetic signals of origin, spread, and introgression in a large sample of maize landraces. *Proceedings of the National Academy of Sciences of the United States of America, 108*, 1088–1092. https://doi.org/10.1073/pnas.1013011108

Heiser, C. B. (1989). Domestication of Cucurbitaceae: *Cucurbita* and *Lagenaria*. In D. R. Harris & G. C. Hillman (Eds.), *Foraging and farming: The evolution of plant exploitation* (pp. 471–480). London: Unwin Hyman.

Hu, Y., Chen, J., Fang, L., Zhang, Z., Ma, W., Niu, Y., . . . Baruch, K. (2019). *Gossypium barbadense* and *Gossypium hirsutum* genomes provide insights into the origin and evolution of allotetraploid cotton. *Nature Genetics, 51*, 739–748. https://doi.org/10.1038/s41588-019-0371-5

Jackson, D. D. (1989). Searching for medicinal wealth in Amazonia. *Smithsonian, 19*, 94–102.

Jenkins, J. A. (1948). The origin of the cultivated tomato. *Economic Botany, 2*, 379–392. https://doi.org/10.1007/BF02859492

Johns, T. (1989). A chemical-ecological model of root and tuber domestication in the Andes. In D. R. Harris & G. C. Hillman (Eds.), *Foraging and farming: The evolution of plant exploitation* (pp. 504–522). London: Unwin Hyman.

Johns, T., & Kubo, I. (1988). A survey of traditional methods employed for the detoxification of plant foods. *Journal of Ethnobiology, 8*, 81–129.

Keatinge, R. W. (1988). *Peruvian prehistory*. Cambridge, UK: Cambridge University Press.

Kraft, K. H., Brown, C., Nabhan, G., Luedeling, E., Luna Ruiz, J., Coppens d'Eeckenbrugge, G., . . . Gepts, P. (2014). Origin of domesticated chili pepper in Mexico. *Proceedings of the National Academy of Sciences of the United States of America, 111*, 6165–6170. https://doi.org/10.1073/pnas.1308933111

Krapovickas, A., & Gregory, W. C. (1994). Taxonomy of the genus *Arachis* (Leguminosae). *Bonplandia, 8*, 1–186.

Lathrap, D. W. (1977). Our father the cayman, our mother the gourd: Spinden revisited, or a unitary model for the emergence of agriculture in the New World. In C. A. Reed (Ed.), *Origins of agriculture* (pp. 713–751). The Hague, The Netherlands: Monton. https://doi.org/10.1515/9783110813487.713

Leitch, I. J., Hanson, L., Lim, K. Y., Kovarik, A., Chase, M. W., Clarkson, J. J., & Leitch, A. R. (2008). The ups and downs of genome size evolution in polyploid species of *Nicotiana* (Solanaceae). *Annals of Botany (London), 101*, 805–814. https://doi.org/10.1093/aob/mcm326

Liu, H., Wang, X., Warburton, M. L., Wen, W., Jin, M., Deng, M., . . . Yan, J. (2015). Genomic, transcriptomic, and phenomic variation reveals the complex adaptation of modern maize breeding. *Molecular Plant, 8*, 871–884. https://doi.org/10.1016/j.molp.2015.01.016

Lynch, T. F. (1980). *Guitarrero Cave: Early man in the Andes*. New York, NY: Academic Press.

Mangelsdorf, P. C., MacNeish, R. S., & Galinat, W. C. (1967). Prehistoric maize, teosinte, and *Tripacum* from Tamaulipas, Mexico. *Botanical Museum Leaflets Harvard University, 22*, 33–63.

Matsuoka, Y., Vigouroux, Y., Goodman, M. M., Sanchez, J., Buckler, E., & Doebley, J. (2002). A single domestication for maize shown by multilocus microsatellite genotyping. *Proceedings of the National Academy of Sciences of the United States of America, 99*, 6080–6084. https://doi.org/10.1073/pnas.052125199

Meggers, B. J., Ayensu, E. S., & Duckworth, D. H. (1973). *Tropical forest ecosystems in Africa and South America: A comparative review*. Washington, DC: Smithsonian Institution Press.

Mutis, J. C. (1957). *Diario de observaciones* (Vol. 1). Bogota, Colombia: Instituto Colombiano de cultura Hispanica, Editorial Minerva, LTDA.

Nabhan, G., & deWet, J. M. J. (1984). *Panicum sonorum* in Sonoran desert agriculture. *Economic Botany, 38*, 65–82. https://doi.org/10.1007/BF02904417

Nishiyama, I. (1971). Evolution and domestication of the sweet potato. *Botanical Magazine, Tokyo, 84*, 377–387. https://doi.org/10.15281/jplantres1887.84.377

Olsen, K. M., & Schaal, B. A. (1999). Evidence on the origin of cassava: Phylogeography of *Manihot esculenta*. *Proceedings of the National Academy of Sciences of the United States of America, 96*, 5586–5591. https://doi.org/10.1073/pnas.96.10.5586

Perry, L., Dickau, R., Zarrillo, S., Holst, I., Pearsall, D. M., Piperno, D. R., . . . Zeidler, J. A. (2007). Starch fossils and the domestication and dispersal of chili peppers (*Capsicum* spp. L.) in the Americas. *Science, 315*, 986–988. https://doi.org/10.1126/science.1136914

Phillips, L. L. (1976). Cotton. In N. W. Simmonds (Ed.), *Evolution of crop plants* (pp. 196–200). London: Longman.

Pickersgill, B. (1989). Cytological and genetical evidence on the domestication and diffusion of crops within the Americas. In D. R. Harris & G. C. Hillman (Eds.), *Foraging and farming: The evolution of plant exploitation* (pp. 426–439). London: Unwin Hyman.

Pickersgill, B. (2007). Domestication of plants in the Americas: Insights from Mendelian and molecular genetics. *Annals of Botany (London), 100*, 925–940. https://doi.org/10.1093/aob/mcm193

Piperno, D. R. (1989). Non-affluent foragers: Resource availability, seasonal shortage, and the emergence of agriculture in Panamanian tropical forests. In D. R. Harris & G. C. Hillman (Eds.), *Foraging and farming: The evolution of plant exploitation* (pp. 538–554). London: Unwin Hyman.

Piperno, D., & Pearsall, D. (1998). *The origins of agriculture in the lowland neotropics*. San Diego, CA: Academic Press.

Ranere, A. J., Piperno, D., Holst, I., Dickau, R., & Iriarte, J. (2009). The cultural and chronological context of early Holocene maize and squash domestication in the Central Balsas River Valley, Mexico. *Proceedings of the National Academy of Sciences of the United States of America, 106*, 5014–5018. https://doi.org/10.1073/pnas.0812590106

Recinos, A. (1947). *Popol Vuh, Las antiquas historias del Quiché. Ciudad de Mexico*. Mexico, North America: Fondo de Cultura Económica.

Rendón-Anaya, M., Montero-Vargas, J. M., Saburido-Álvarez, S., Vlasova, A., Capella-Gutierrez, S., Ordaz-Ortiz, J. J., . . . Herrera-Estrella, A. (2017). Genomic history of the origin and domestication of common bean unveils its closest sister species. *Genome Biology, 18*, 60. https://doi.org/10.1186/s13059-017-1190-6

Reyna, R. M., & Alvarez, C. (2009). El maíz arqueológico, evidencia ancestral de las razas de maíz contemporáneas: El caso de La Organela Xochipala, Guerrero. In Coordinación Nacional de Antropología del Instituto Nacional de Antropología e Historiam (ed), In *Diario de campo, desgranado una mazorca, orígenes y etnografía de los maíces nativos* (Vol. *52*, pp. 28–39). México, DF: Publicación interna de la Coordinación Nacional de Antropología del Instituto Nacional de Antropología e Historiam.

Rick, J. W. (1980). *Prehistoric hunters of the high Andes*. New York, NY: Academic Press.

Riley, C. L., Kelley, J. C., Pennington, C. W., & Rands, R. L. (1971). *Man across the sea*. Austin, TX: University of Texas Press.

Ross-Ibarra, J., Tenaillon, M., & Gaut, B. S. (2009). Historical divergence and gene flow in the genus *Zea. Genetics, 181*, 1399–1413. https://doi.org/10.1534/genetics.108.097238

Roullier, C., Duputié, A., Wennekes, P., Benoit, L., Bringas, V. M. F., Rossel, G., . . . Lebot, V. (2013a). Disentangling the origins of cultivated sweet potato (*Ipomoea batatas* (L.) Lam.). *PLoS One, 8*(5). https://doi.org/10.1371/journal.pone.0062707

Roullier, C., Duputié, A., Wennekes, P., Benoit, L., Bringas, V. M. F., Rossel, G., . . . Lebot, V. (2013b). Correction: Disentangling the origins of cultivated sweet potato (*Ipomoea batatas* (L.) Lam.). *PLoS One, 8*(10). https://doi.org/10.1371/annotation/936fe9b4-41cb-494d-87a3-a6d9a37c6c68

Schmitz, P. I. (1987). Prehistoric hunters and gatherers of Brazil. *Journal of World Prehistory, 1*, 53–126. https://doi.org/10.1007/BF00974817

Schultes, R. E., & Hofmann, A. (1979). *Plants of the gods: Origins of halucinogenic use.* New York, NY: McGraw-Hill Book Co.

Shipek, F. C. (1989). An example of intensive plant husbandry: The Kumeyaay of southern California. In D. R. Harris & G. C. Hillman (Eds.), *Foraging and farming: The evolution of plant exploitation* (pp. 159–170). London: Unwin Hyman.

Shutler, R. (1983). *Early man in the New World.* Beverly Hills, CA: Sage Publications.

Spooner, D. M., McLean, K., Ramsay, G., Waugh, R., & Bryan, G. J. (2005). A single domestication for potato based on multilocus amplified fragment length polymorphism genotyping. *Proceedings of the National Academy of Sciences of the United States of America, 102*, 14694–14699. https://doi.org/10.1073/pnas.0507400102

Stalker, H. T., & Simpson, C. E. (1995). Germplasm resources in *Arachis*. In H. E. Pattee & H. T. Stalker (Eds.), *Advances in peanut science* (pp. 14–53). Stillwater, OK: American Peanut Research and Education Society.

Stanford, D. (1983). Pre-Clovis occupation south of the ice sheets. In R. Shutler (Ed.), *Early man in the New World* (pp. 65–72). Beverly Hills, CA: Sage Publications.

Tian, F., Stevens, N. M., & Buckler, E. S. (2009). Tracking footprints of maize domestication and evidence for a massive selective sweep on chromosome 10. *Proceedings of the National Academy of Sciences of the United States of America, 106*(Supplement 1), 9979–9986. https://doi.org/10.1073/pnas.0901122106

Warburton, M. L., Wilkes, G., Taba, S., Charcosset, A., Mir, C., Dumas, F., ... Franco, J. (2011). Gene flow among different teosinte taxa and into the domesticated maize gene pool. *Genetic Resources and Crop Evolution, 58*, 1243–1261. https://doi.org/10.1007/s10722-010-9658-1

Watson, P. J. (1989). Early plant cultivation in the eastern woodlands of North America. In D. R. Harris & G. C. Hillman (Eds.), *Foraging and farming: The evolution of plant exploitation* (pp. 555–571). London: Unwin Hyman.

Wendel, J. F., Brubaker, C. L., & Percival, A. E. (1992). Genetic diversity in *Gossypium hirsutum* and the origin of upland cotton. *American Journal of Botany, 79*, 1291–1310. https://doi.org/10.1002/j.1537-2197.1992.tb13734.x

Wilkes, H. G. (1967). *Teosinte: The closest relative of maize.* Cambridge, MA: The Bussey Institute, Harvard University.

Wycherley, P. R., & Simmonds, N. W. (1976). Rubber. In *Evolution of crop plants* (pp. 77–81). London: Longman.

Wynne, J. C., & Coffelt, T. A. (1982). Genetics of *Arachis hypogaea* L. In H. E. Pattee & C. T. Young (Eds.), *Peanut science and technology* (pp. 50–94). Yoakum, TX: American Peanut Research and Education Society.

Yen, D. E. (1974). *The sweet potato and Oceania. Bernice P. Bishop Museum Bulletin, 236*. Honolulu, HI: Bernice P. Bishop Museum.

Zhang, T., Hu, Y., Jiang, W., Fang, L., Guan, X., Chen, J., . . . Chen, Z. J. (2015). Sequencing of allotetraploid cotton (*Gossypium hirsutum* L. acc. TM-1) provides a resource for fiber improvement. *Nature Biotechnology, 33*, 531–537. https://doi.org/10.1038/nbt.3207

12

Epilogue

Who's in Charge Here?

It is as if man had been appointed managing director of the biggest business of all, the business of evolution . . . whether he is conscious of what he is doing or not he is in point of fact determining the future direction of evolution on this earth. That is his inescapable destiny, and the sooner he realizes it and starts believing in it the better for all concerned.

Julian S. Huxley (1957)

Our best information, then, indicates that many small, tentative steps toward food production were undertaken, widely scattered over the earth in both the Old and New Worlds. Some of these prospered, some aborted, and some remained tentative up to the present. In due time, the more

Source: Patricia J. Scullion

Harlan's Crops and Man: People, Plants and Their Domestication, Third Edition.
H. Thomas Stalker, Marilyn L. Warburton, and Jack R. Harlan.
© 2021 American Society of Agronomy, Inc. and Crop Science Society of America, Inc.
Published 2021 by John Wiley & Sons, Inc.
doi:10.2135/harlancrops

successful attempts began to network and interact, and the process intensified. A few millennia later, fully developed agricultural systems had evolved sufficiently to support urban civilizations. The world moved into ancient history and eventually into modern times. Today, agriculture provides food, fiber, ornamentals, industrial products, and many medicines for more than 7.8 billion people (as of 2020), and will support many more in the current century.

Of the thousands of plant species used by hunter-gatherers and traditional farmers, only a very few are produced in sufficient quantities to have much impact on the human diet today. The process of discarding crops still goes on. Vegetables that were fairly common in the early 1900s can hardly be found at all today. In the United States, no more than five or six varieties of any kind of fruit appear in the supermarkets. More diversity can be found by diligent search in specialty stores and farmer's markets, but the choices are diminishing year by year. As the world population increases, the food supply depends more and more on fewer and fewer species and less and less on diversity of species in production. On the other hand, with the rise of the local and "foodie" cultures in many developed countries, interest in traditional or heirloom varieties of many vegetables, fruits, nuts, legumes, and even cereals is slightly reversing the trend of diversity loss. Seed catalogs and seed saver's networks now complement the public gene banks as reservoirs and sources of genetic diversity for some species, at least. However, this is only a drop in the bucket of lost diversity.

The percentage of people engaged full time in agricultural production has declined drastically in developed countries. At the emergence of civilizations in Mesopotamia, Egypt, China, and others, were about 80% of the people were engaged directly in food production. This permitted some 20% of the population to go into professions, trade, manufacture, standing armies, priestly castes, and so on. The ratio remained more or less constant until the industrial revolution permitted some mechanization of agriculture. Steam tractors and threshing machines had some effect, but it was mass production of gasoline and diesel engines that made modern mechanization feasible. Today, less than 2% of the population produces the crops in most industrialized countries. Perhaps the best measure of degree of "development" of a country is percent of the population devoted to full-time farming. The smaller the figure, the greater the industrialization. Enormous urban populations are now sustained by a relative handful of producers.

The great concern, of course, is some breakdown in the system, resulting in food shortage, famine, and starvation. Serious famines seem to have come with the development of agriculture. Studies of human skeletons and bones have revealed that the health status of agricultural people was not nearly as good as that of hunter-gatherers, but outright famine is of a different order. Archaeology and history have both shown that famine has been a constant companion of agricultural societies. We mentioned the first intermediate period of Egypt when the Old Kingdom collapsed due to low floods on the Nile. Could it happen again? Of course; it has happened repeatedly to various degrees. A second intermediate period destroyed the Middle Kingdom about 1000 years later, circa 3200 BP. The misery was compounded by invasion of the Sea People, but these, whoever they were, were probably driven from their homeland by drought.

Arabian records compiled by Toussoun (1925) show low floods between 944 and 971 AD, and again in 997 AD with famine following in 1007 and 1008 AD. Disastrous low floods occurred between 1065 and 1072 AD; there were other brief periods of serious low water. Around 9–12 million people died in China in 1876–1879 and another 3 million in 1928–1929 due to crop loss after floods. Central and western Europe suffered grotesquely between 1315 and 1317 due to a combination of too much rain and plagues. Fields were too wet to work, or if crops were planted, they did not mature or rotted before harvest. Disease always erupts in populations weakened by starvation, but the Black Death was particularly virulent. The European population as a whole was reduced by about 10%; locally the death toll reached as high as 90% and many villages disappeared entirely. These things have happened with regularity and will happen again. In fact, they seem to happen somewhere in the world in any given year, and this despite modern flood prevention infrastructure in much of the world. In 2018, for example, floods in Eastern Africa caused widespread crop loss only one year after drought caused food shortages in the same region.

A lack of snow in the Sierra Nevada and Cascade ranges of our western states could be considered the American equivalent of low floods on the Nile. Unfortunately, this now happens with disturbing regularity. Everyone in California, urban and rural, depends in one way or another on snow in the mountains and water delivered to homes and farms downslope. Suppose there was no snow for six consecutive years? What farms would be left? These things do happen and can happen again.

Historically, famines have been mostly local, rarely affecting a whole continent, and more often than not with a social and political component. During the Bengal famine of 1943–1944, an estimated 1.5 million people died of starvation. The trouble was initiated by poor rice harvests due to blast disease, but the real damage was done by high prices that put rice out of reach of the poor. The great famine in the former Soviet Union in 1932–1933 caused an estimated 5 million deaths due to Stalin's drive to collectivize the farming enterprise. A drought in the Sahel region led to 1 million deaths in Mali, Chad, Niger, Mauritania, and Burkina Faso in the 1970s; during this period, internal instability was both cause and effect of the famine in these countries. Even the famous potato famines of Ireland 1846–1847 and on to 1851, had sociopolitical components. True, the potato crop was virtually a total loss due to late blight, but the grain crop would have been adequate had it not been confiscated. Some 2 million people died and another million or more emigrated to the United States with profound effects on Boston politics, the New York police force, and the composition of Notre Dame football teams. Famine has been used as a weapon against inhabitants of Eritrea, Somalia, and Sudan because of their efforts to achieve independence.

All threats to the food supply today, including diseases and pests, mismanagement of resources, soil erosion, and a changing climate, are compounded by the problem of too many people, and the populations of the poorest and most vulnerable countries are the ones rising at the fastest rates. Populations in the industrialized countries are approaching or have achieved stability, but we can expect populations in a number of developing countries to outstrip the food supply from time to time. For the world at large, an adequate food supply depends on the performance of the top one-third or so of the select 30 crops (Table 2.1). Here the trend toward uniformity could be disastrous. Fortunately, the people most involved with the genetic manipulation of these crops are well aware of the hazards. Germplasm can be well managed, and current concern over loss of diversity is encouraging. However, being aware of the problem is not necessarily enough. Disasters can be generated with the best of intentions. We have not had a very good track record in playing God. A few examples may illustrate the state of our competence.

Traditionally, crown rust has been the most serious disease of oats in the midwestern United States. The cultivar "Victoria" was a welcome introduction from Uruguay, because it had superior resistance to the disease. The "Victoria" resistance was soon incorporated into several cultivars which were widely deployed throughout the region. They all collapsed due to a disease organism so obscure it had never been described or named. The causal agent was named *Helminthosporium victoriae* (changed to *Cochliobolus victoriae* R.R. Nelson) in honor of the cultivar that lifted it from obscurity. It previously had been a minor disease of native grasses. How could one have predicted this?

This is, by no means, the only case. Pitch canker *(Fusarium moliniforme)* of pine was such a minor disease that it was not described until 1946. Under plantation management, it has become an extremely serious disease of slash pine (*Pinus elliottii* Engelm.) in Florida. Fusiform rust, little leaf disease, and brown spot needle blight cause little or no damage in natural stands, but are destructive in planted forests.

Some of our errors have been more predictable. It has been a common practice, in past years, for farmers in the southeastern United States to clear land from forest, farm for a few years, then abandon the fields, often for economic reasons. More recently, land is cleared and trees sold, then allowed to regenerate. On the Coastal Plain especially, the loblolly pine, *Pinus taeda* L., is the first tree to become established in old field succession. It is well adapted and grows quickly, so it seemed reasonable to plant trees in old fields rather than wait for natural populations to develop. To obtain seeds, most forest agencies and land owners advertised that they would buy seed by the bushel. Originally, it was thought that such seeds would represent natural populations. With some thought, one might have realized that seeds obtained by this procedure would come from runty, dwarfed, easy-to-climb trees—and they did. Untold hectares were planted to genetic runts that would assure slow growth and low production (Dorman, 1976). Fortunately, forestry in the southern United States is now much improved and newer stands are much more productive.

The Colorado potato beetle was once confined to the eastern slope of the Rocky Mountains, where it fed on native species of *Solanum* and did little damage. When settlers began to grow potatoes within its territory, it changed its feeding habits. It spread over North America, found its way to Europe and other potato-growing regions. In its travels, it found other *Solanaceae* to its liking and began to feed on tobacco, peppers, eggplants, and other crops. This obscure insect from Colorado was raised from insignificance to the position of world traveler and pest of distinction. This has happened with countless other introduced pests as well, as globalization marches relentlessly forward.

Our record of accomplishment on use of insecticides is not likely to win any blue ribbons. The gypsy moth spread farther and faster under spray than it ever did without sprays. Resistance to insecticides increased frequency and dosage of spraying cotton. Control of the imported fire ant has not worked to date. Suppression of one species almost always results in irruption of another species. High hopes for eliminating malaria have been dashed by resistant mosquitoes. In fact, Garcia and Huffaker (1979) report that efforts at malaria control in Bolivia resulted in killing household cats with a consequent increase in rodents and outbreaks of hemorrhagic fever. Beyond this, we have yet to measure the consequences of the enormous tonnages of poisons put into the environment that have not achieved expected results. We know something of the enormous damage to wildlife caused by DDT and have observed the recovery since it was banned;

however, many other poisons have yet to be assessed. Integrated pest management is supposed to be the new approach, but often we do not have enough information to implement a program. The trend is in the right direction, however, and we will continue to learn through experience.

We have done better with herbicides, which are generally not so deadly to animals, but spread of herbicide-resistant "Superweeds" have begun to limit the effectiveness of the herbicides. To date, 23 of 26 herbicide sites of action have been overcome by previously controlled weed species. This includes 167 herbicides that are used on 93 crops in 70 countries, and the numbers increase yearly (www.weedscience.org). Questions have been raised about the carcinogenic properties of the chemicals and their breakdown products. The effects of Agent Orange in Vietnam is still under dispute, but cancers and birth defects are increasingly attributed to this chemical. Even the more innocuous chemicals, such as glyphosate (Roundup) are coming under closer scrutiny, and while they have been repeatedly proven to be less toxic than most herbicide classes, they will probably be phased out due to resistant superweeds and consumer backlash. Unfortunately, glyphosate will probably be replaced by something more toxic, as farming without herbicides is exceedingly difficult and costly.

The use of natural allelic diversity for host plant resistance to biotic stresses is particularly promising, and examples of one or a few genes introduced from wild relatives or landrace varieties are common. Resistance to abiotic stresses is often more difficult, as these resistance mechanisms tend to be controlled by many genes, all of which have to be transferred by backcrossing to modern cultivars, without bringing in other traits causing weediness, lower yield, or other agronomic problems in the process. The introduction of one or two genes into the germline of a species, often from very unrelated organisms, via transgenics, to create genetically modified organisms, has been commercially very successful for a few traits, but not for most of the serious problems affecting modern agriculture.

The promise of newer biotechnology tools may increase the impact of modern science. The use of genomic selection allows computer algorithms to predict which plants will be the most productive for any given trait, often without having to measure those traits on all the plants in the breeding population, saving considerable time and expense during cultivar development. Finally, new tools for gene editing will allow single genes to be changed within a species, without having to introduce any genetic material from other species, to create plants, animals, insects, fungi, or bacteria with a new, beneficial trait. Gene editing of humans is as easily possible to correct genetic defects . . . or create new genetic traits. The debate as the usefulness and mortality of these tools has only just started.

Regardless of how the genes of a crop plant are modified—via crossing to landraces, transgenesis of genes from bacteria or fish, manipulation of single base pairs to precisely correct or change the trait—knowledge about the underlying genes is useful, and the more advanced the technology being used, the more necessary this knowledge becomes. Knowing how to tweak a gene is not useful if you don't know which gene to tweak, or how to tweak it to cause the expression of a useful trait. You can make random changes to genes and look at the outcome; you can even change every gene in an organism's genome, one at a time, and grow out the new plant to see what that change did. But these random or individual changes will not be useful, or even cause a visible difference, in the vast majority of the resulting edited plants grown under most environmental conditions. For example, a change in a plant that makes it more disease resistant will only be seen when the plant is grown with the disease agent. And if you need changes in two genes simultaneously, you will never find them both by testing one gene at a time. You would already have to know what the gene in question was likely to affect to know how best to test the plant that you had edited for that gene.

The information needed to determine what each gene in a plant's genome is doing resides within the vast collections of plant genetic resources in the world's gene banks. The landraces, wild relatives, and genetic stocks held within these gene banks have the variation within the sequences of all the genes of each species. When these are studied in large trials with groups of several hundred diverse individuals, nearly endless combinations of genetic sequence differences can be studied and their effect on traits of interest determined. Once useful changes to the DNA are identified for these useful traits, they can be repeated, and improved upon in the laboratory. Science promises that big data will improve our lives; the big data currently encoded for in the billions of base pairs in the genomes of the millions of seeds in the gene banks is what is needed. However, this is hardly feasible for the vast majority of species.

If we have learned anything, we should have learned by now that simple solutions simply do not work very well. Our only hope for achieving stability and sustainability in our managed ecosystems is to imitate natural ecosystems as much as possible. We have a lot to learn by studying natural defenses against diseases and pests. We can learn more by studying the genetic architecture of landraces that have evolved over centuries and millennia. We can learn much from traditional farmers who have developed the arts of survival. We need to approach the daunting tasks ahead with more humility and take a broader view of the ecosystems we must manage. Technology alone is not the answer. The human is a clever animal and can learn from mistakes, but can we learn enough fast enough to avoid making colossal blunders? We are now in a position where we must not only manage our crop plants, our domestic animals, our fisheries, our forests and range lands, but the whole globe is in our care, ready or not, competent or not. We are affecting the atmosphere, the oceans, the forests, rainforests, deserts, and even the climate. We are woefully unprepared for this awesome responsibility. This is an age of great knowledge and little wisdom, but we have no choice; we must blunder on. Who is in charge here? God help us, we are!

References

Dorman, K. W. (1976). *The genetics and breeding of southern pines. USDA Handbook No. 471*. Washington, DC: U.S. Department of Agriculture, Forest Service, U.S. Government Printing Office.

Garcia, R., & Huffaker, C. B. (1979). Ecosystem management for suppression of vectors of human malaria and schistosomiasis. *Agro-Ecosystems, 5*(4), 295–315. https://doi.org/10.1016/0304-3746(79)90033-7

Huxley, J. S. (1957). *New bottles for new wine*. New York, NY: Harper & Bros.

Toussoun, O. (1925). *Mémoire sur l'histoire du Nil*. Cairo, Egypt: Imprimerie de l'Institut Français d'archéologie Orientale.

Printed and bound by CPI Group (UK) Ltd, Croydon, CR0 4YY

20/11/2023

08191581-0004